The Future
of Packaging

Reflections on *The Future of Packaging*

"Plastics came of age in the 1950s, changing manufacturing forever. By telling the story that leads us to today's linear packaging model, I illustrate that designing into the circular systems that came before it can be a short journey back."

—**Attila Turos, Former Lead, Future of Production Initiative, World Economic Forum**

"In *The Future of Packaging*, we talk about the modern problem of waste, how packaging fits into that, and how we can design out of it. It is important to explain the forces that catalyzed the first formal recycling programs in the United States, defining the need to scale up on today's systems."

—**Christine "Christie" Todd Whitman, President, The Whitman Strategy Group; Former Governor of New Jersey; Former Administrator, Environmental Protection Agency**

"Moving away from the linear take-make-waste model is an ethical imperative. In my chapter I talk about the fragmented global recycling system and how investing in it presents opportunity for innovation, jobs creation, education, and, above all, prosperity."

—**Jean-Marc Boursier, Group Senior Executive Vice President and Chief Financial Officer, Finance and Recycling & Recovery (Northern Europe), SUEZ**

"Whether you are a packaging manufacturer, small business, local government, or consumer, this book will transform the issues we've avoided into ones we are motivated to tackle head-on. My chapter calls for a paradigm shift in producer responsibility, placing waste and materials management in the hands of the producer as an asset, not a burden."

—**Scott Cassel, Founder and CEO, Product Stewardship Institute**

"Almost everything is technically recyclable, so why do we have so much waste? Improving our recycling system will help us turn more waste into worth. When we view recycling in terms of supply and demand, it is much easier to see where system advancements are needed. We hope *The Future of Packaging* brings this to life and shows how all of us can do our part to keep our environment and oceans free from litter."

—**Stephen Sikra, Associate Director, Corporate R&D, Procter & Gamble**

The Future of Packaging

From Linear to Circular

Tom Szaky

and 15 Industry
Leaders in Innovation
and Sustainability

BK

Berrett–Koehler Publishers, Inc.

Berrett-Koehler Publishers, Inc.
1333 Broadway, Suite 1000, Oakland, CA 94612-1921
Tel: (510) 817-2277   Fax: (510) 817-2278   www.bkconnection.com

Ordering Information

Quantity sales. Special discounts are available on quantity purchases by corporations, associations, and others. For details, contact the "Special Sales Department" at the Berrett-Koehler address above.

Individual sales. Berrett-Koehler publications are available through most bookstores. They can also be ordered directly from Berrett-Koehler: Tel: (800) 929-2929; Fax: (802) 864-7626; www.bkconnection.com.

Orders for college textbook/course adoption use. Please contact Berrett-Koehler: Tel: (800) 929-2929; Fax: (802) 864-7626.

Distributed to the U.S. trade and internationally by Penguin Random House Publisher Services.

Berrett-Koehler and the BK logo are registered trademarks of Berrett-Koehler Publishers, Inc.

Printed in the United States of America

Berrett-Koehler books are printed on long-lasting acid-free paper. When it is available, we choose paper that has been manufactured by environmentally responsible processes. These may include using trees grown in sustainable forests, incorporating recycled paper, minimizing chlorine in bleaching, or recycling the energy produced at the paper mill.

Library of Congress Cataloging-in-Publication Data
Names: Szaky, Tom. Future of packaging.
Title: The future of packaging : from linear to circular / Tom Szaky and
   15 industry leaders in innovation and sustainability.
Description: 1st Edition. | Oakland, California : Berrett-Koehler Publishers,
   [2019]
Identifiers: LCCN 2018037199 | ISBN 9781523095506 (paperback)
Subjects: LCSH: Package goods industry—Environmental aspects. |
   Packaging waste—Environmental aspects. | Sustainable development.
   | BISAC: HOUSE & HOME / Sustainable Living. | BUSINESS &
   ECONOMICS / Green Business. | TECHNOLOGY & ENGINEERING /
   Environmental / Waste Management.
Classification: LCC TD195.P26 F88 2018 | DDC 688.8068/4—dc23
LC record available at https://lccn.loc.gov/2018037199

24 23 22 21 20 19 18   10 9 8 7 6 5 4 3 2 1

Cover design by Dan Tesser, Studio Carnelian. Interior design and composition by Gary Palmatier, Ideas to Images. Elizabeth von Radics, copyeditor; Mike Mollett, proofreader; Paula Durbin-Westby, indexer.

In memory of
Martin Stein and Robin Tator

Contents

CHAPTER **14**

CHAPTER **15**

Foreword

Paul Polman
CEO, Unilever

I T IS AN UNCOMFORTABLE TRUTH THAT MUCH OF BUSINESS growth has been founded on unsustainable models of production and consumption. About 60 percent of the world's resources are already degraded or used unsustainably, and the earth has passed only minimal acceptance levels for the critical life-support systems vital for human survival, including biodiversity, fresh water, and land use.[1] We have a rising middle class, a world population projected to reach 9 billion by 2050,[2] and a planet approaching the limit of its ability to provide.

A continuation of business as usual would mean not just a slight additional strain but an inevitable crisis. The Ellen MacArthur Foundation found that 95 percent of the value of plastic packaging on the market is lost after its first use. That is equivalent to $80 billion to $120 billion of lost profit.[3] If we don't change, we will experience a status quo economy of decline—and worse.

At Unilever we understand that we are doing business in an age of mounting environmental, political, and social problems tied to the current one-way models, and we recognize that things must change. That is why we created the Unilever

Sustainable Living Plan: to decouple growth from environmental impact with a total-value-chain approach to growing business. Six years in, we are 80 percent on track for more than 50 targets, including greening our energy use toward carbon positivity by 2030 and making all of our packaging either fully compostable or recyclable by 2025.

We continue to examine business models in a new light, designing plans and products for increased sustainability and a lower carbon footprint. Just last year we piloted a new technology to recycle *sachets*—those small, single-use, plastic pouch–like items used to package household goods, personal care items, and food products that are largely unrecyclable and end up littered, landfilled, and heavily polluting. Now sachets can be turned into plastic pellets that can be used in the manufacture of new packaging and channeled back into the supply chain.

A complete overhaul of the way we produce and consume is needed, and making resource-recovery technologies and regenerative production systems an open source for the broader industry is key to wider systems change. The journey out of the current one-way, make-use-dispose *linear economy* that views products and packaging as disposable after one use will not be an easy one. We need to work up to and into a *circular economy*—the make-use-recycle-remanufacture concept wherein all materials are kept at high utility and "waste" outputs are useful inputs in the production process.

The changes needed to achieve this ideal require authentic commitments from a variety of stakeholders and, more than that, boldness. Players in all facets of consumer goods production and consumption need to collaborate to reexamine what we create—and take responsibility for it. We need more than one innovation, and we need ones that will stick, which makes the publication of this book so timely.

The Future of Packaging: From Linear to Circular is a crash course on designing for the circular economy. The book is steered by Tom Szaky, waste pioneer, eco-capitalist, and founder and CEO of TerraCycle, and each chapter is co-authored by an expert in their field. From the distinct perspectives of government leaders, consumer packaged goods companies, waste management firms, and more, the book explores current issues of production and consumption, practical steps for improving packaging and reducing waste today, and big ideas and concepts we can carry forward tomorrow.

Designed to help everyone, from a small entrepreneurial start-up to a large established consumer products company, move toward a circular economy, the book can be used as a source of knowledge and inspiration. The message from our innovators is not to scale back but to innovate upward to nurture an optimistic vision of a future of abundance and prosperity, with less waste.

Look outside your own silo for new ideas and cast a net as far and wide as possible to get new direction from innovators, designers, entrepreneurs, and scientists around the world. Better solutions are possible only if all the players do their part. This book is for *you*—the future leaders—to engage and inspire you to learn from the best.

From Linear to Circular

Tom Szaky

Founder and CEO, TerraCycle

BY NOW THE OFT-CITED PROJECTION THAT WE WILL SEE more plastics by weight than fish in the ocean by 2050[1] has long spread from its origins at the World Economic Forum in early 2017 to headlines in mainstream news outlets. The public now knows that plastic pollution isn't just the stray piece of litter on the hiking trail or the rogue water bottle on the beach: it's a global crisis, unsightly and largely unseen, and one that affects us all.

Plastic pollution is no longer an issue out of sight and out of mind. News outlets recently reported that 94 percent of tap water samples collected in the United States tested positive for *microplastic*[2]—those tiny, often microscopic particles that form when plastics break down into smaller particle sizes in the environment. Plastic waste contaminates both our water and our food,[3] leaching chemicals and heavy metals in our water sources and the ground where we grow our crops and, in the end, bioaccumulating in our bodies through the food we eat and the water we drink. The Centers for Disease Control and Prevention recently reported that cancer rates in men and women are projected to increase by 24 percent and 21 percent, respectively,[4] between 2010 and 2020, ironically all while our cost of living has increased at unprecedented rates.

The more we look, the more we find. Just last year an uninhabited island in the South Pacific, so remote that the nearest human settlement is a small island 200 kilometers away, was recorded to be covered by more than 35 million pieces of plastic. Chunks of plastic discovered on the ice and floating in the oceans of the Arctic had initially prompted fear in researchers about the far reach of this man-made waste. Then it was discovered that the levels of plastic pollution in Antarctica—the very end of the earth—were five times worse than predicted. **SEE I.1**

What is just now surfacing as a revelation to consumers has been long known to manufacturers, retailers, and governments—and confirmed and obscured by the plastics industry for much longer: plastics are nearly indestructible, far from "disposable," and highly polluting.[5]

The Current State of Affairs

Only 35 percent of the 240 million metric tons of waste generated in the United States per year gets recycled.[6] Nearly half a century has passed since the launch of the first universal recycling symbol, and recent estimates say that 91 percent of plastic is not recycled.[7] Plastics production is outpacing waste management, let alone recycling, and a significant portion of

DutchScenery/Shutterstock

I.1 Plastic pollution is now recorded as having reached the most remote and pristine places on earth.

the problem comes down to our addiction to single-use packaging and disposable products.

To effectively meet this challenge, we need to understand its origins and why our world is addicted to disposability. And there is room for hope: the idea of waste is modern—roughly 70 years old. It was only in the 1950s that complex materials became commercialized on a mass scale, replacing age-old reusable models like "the milkman" and cobbling one's shoes. They did so because single-use, "disposable" products and packages make life more convenient and affordable, increasing access to goods for all strata of society. As a result, humanity makes and purchases 70 times more stuff today than we did in the 1950s, stuff that is typically made from materials that nature lacks the capacity to absorb.

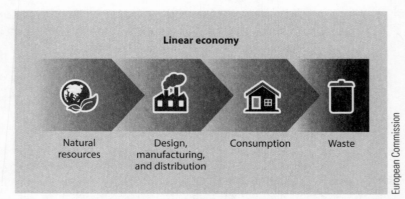

I.2 The linear economy is a one-way, take-make-waste resource model that tracks the world's limited pool of resources to landfilling and incineration. Why?

Today we don't mend our clothing and refill our beer bottles. Instead of buying high-quality *durable goods* that tend to be more expensive, we buy lots of cheaper disposable goods, as they give us immediate happiness and utility, even if they don't last. The trade-off is that most packaging is discarded as soon as a product is opened or used. In fact, 99 percent of all stuff becomes waste within the first 12 months of purchase. As producers, industry knowingly and continually creates products and packaging on a one-way track to landfilling and incineration—buried or burned. This make-use-dispose pipeline has become known as the *linear economy* because products and packaging, once manufactured and used, too often go in one direction: the garbage. **SEE I.2**

This trend has become only more pronounced as manufacturers are able to produce packaging at a fraction of the cost of their predecessors. If we bought juice 70 years ago, it

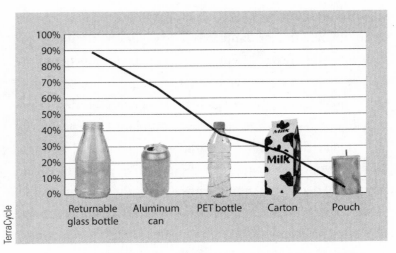

TerraCycle

I.3 Every "innovation" that has made product packaging lighter, more convenient, and less expensive to make has effectively cut its recyclability in half.

was in a returnable glass bottle, which in the United States was reused at more than 93 percent. Since then we have moved to aluminum cans (recycled at 67 percent); to *polyethylene tereph-thalate* (**PET**, or #1 plastics) bottles, which are recycled at half the rate (32 percent); to cartons, which are recycled at half the rate of plastic; and, finally, to **pouches**, which are not recycled municipally at all. While this may seem deplorable, it's good for profit: a pouch is 97 percent less material than a heavy-duty refillable glass bottle, making it phenomenally cheaper per liter of juice packaged and requiring less material use. **SEE I.3**

Designing out of Waste

Luckily, the very idea of waste is relatively modern, as is our dependence on disposable packaging. Waste-free wisdom is all around us. It's embodied in nature (where the idea of waste is completely foreign) and in our past (where durability and frugality were prized), and it will likely be at the core of our future. If we can learn where the waste problem came from and understand what makes it tick, we can then design our way out of it—all while increasing abundance and prosperity.

Every day designers, purchasers, producers, retailers, and everyone else involved in the production of goods make important decisions that influence the way the world consumes and the footprint we as consumers leave behind. We can choose to design packaging that is more easily recycled in our current systems, or we can establish new packaging platforms where perhaps the package is constantly reused and has a higher purpose than its original intent.

Reconciling innovation and growth with sustainability is by no means an easy task. It is a complex issue that the biggest manufacturers in the world struggle to get right. Recently, I witnessed a senior executive at a global cosmetics brand become genuinely mortified to learn that the black plastic of one of its cosmetics lines (locked into that specific design for five years) is unrecyclable. Even though it's made from PET— the most recycled type of plastic in the world—black plastic is not detectable in recycling centers. Many "green" packaging trends, including weight reduction and the use of naturally *biodegradable* content, are not automatic slam dunks for sustainability and are worthy of a closer look.

Packaging design for profitability maximization is certainly complex enough without considering the full life cycle of materials and their impacts on the supply chain and the environment. No matter your place in the ranks of affecting what is made, from being an individual consumer to the CEO of a major global retailer, it is not easy to champion change. First, we must open our eyes, learn, and understand, and only once we're empowered can we start making and advocating for the right decisions. We all have influence on our future, as what is available in stores is truly a reflection of our current desires and industries—and a best guess at our future ones.

We hope that this book can help everyone—from the individual consumer to the small start-up to the large consumer products company—move toward a world where packaging is not seen as a negative but rather as a fully restorative positive, just like the packages (or skins) that have protected and preserved our fruits and vegetables for millennia.

This book is designed to be a starting point on packaging design for the *circular economy*, wherein we keep resources in use for as long as possible, extract their maximum value while in use, and then recover and regenerate products and materials at the end of each service life to reintegrate them into the supply chain. I have had the privilege to co-author it with 15 of the best minds in the global packaging movement—folks who have been championing this new frame of thinking for decades and in many cases making amazing changes along the way. Speaking from the helm of a company on a mission to recycle the unrecyclable,[8] it is important that we acknowledge the tall order of changing course by addressing it from several angles at the same time.

Thinking critically about the function of packaging, how it fits in the resource economy, and ways we can change the paradigm around consumption is essential to designing our way out of waste and into abundance. No shift is too small, and no innovation is too big.

Plastic, Packaging, and the Linear Economy

Attila Turos

Former Lead, Future of Production
Initiative, World Economic Forum

I MAGINING A WORLD WITHOUT PLASTIC IS NEARLY IMPOS-sible. We interact with it from the moment our digital clocks, smartphones, smart speakers, and Wi-Fi-enabled coffee makers wake us up, to the time we sleep on memory foam mattresses and microfiber sheets. Paper coffee cups are lined with it, razor blades are now forged of it, and lifesaving medicines and treatments are administered and delivered by it in more configurations than ever before. Industries once dominated by metal and other naturally occurring materials (like wood and cotton) have been taken over by plastic, which now makes up roughly 15 percent of the average car by weight and about 50 percent of jets like the Boeing Dreamliner.[1]

Consumers are often surprised to learn just how pervasive plastic is across the entire economy. The corkboard you have hanging on your wall? Cork-colored plastic. Your kid's synthetic fur plush toys and stuffed animals? Plastic. The core of your "wood" door is made of *polyvinyl chloride* (**PVC**) and insulating plastic foam. The textiles of nearly every item in most of our closets are majority oil-based fiber, and very few of us are actually "burning rubber" with our vehicles when we speed off to our next appointment.

Business is largely responsible for this shift. In fact, one of the first man-made plastics was the result of a commission to find an alternative material for school blackboards in the late 1800s,[2] an anecdote illustrative of industry's close ties to the development of plastic as the favored medium for business. Items once carved out of a solid block of wood, forged of steel, or spun out of wool can be more easily made from plastic, which is lighter, stronger, and less expensive to produce, an aspect that has numerous functional, aesthetic, and economic advantages for both companies and consumers.

Modern life now is dependent on the fossil fuel by-product, as the American Petroleum Institute's 2017 "Power Past Impossible" Super Bowl ad reminded the public.[3] The ad shows a robotic prosthesis pulling an arrow firmly back in a bow to reveal itself attached to a young woman. High-contrast blue and magenta of an electrocardiogram displays the beat of an artificial but fully pumping heart valve. An image of a rocky golden landscape is reflected in the helmet of an astronaut who, backlit by fog and sparks, walks away from us, with a pack sporting a decal of the American flag.

A Material of Substance

The spirit of these forward-thinking innovations can be traced back to the discovery and inspired use of natural, bioderived substances such as rubber, egg, and blood proteins by ancient artisans and craftsman (manufacturers in their own right) as early as 1600 BCE.[4] Cutting-edge for their time, the useful behavior of these plasticlike compounds sealed the roofs of dwellings, made containers and pots, and banded goods together for transport, offering an alternative construction material for the business activities of early humans. Since then, plastics have evolved into myriad man-made material types,[5] poised to address changing needs, as well as gaps, in a competitive market.

Synthetic polymers have been disrupting commodity industries for well over 100 years. John Wesley Hyatt patented Celluloid in 1869,[6] a commercially viable solid, stable nitrocellulose used to make things like billiard balls, false teeth, combs, jewelry, and piano keys; it had a comparable performance and look, a more secure supply chain, and a much better price point than the more expensive conventional materials. Celluloid could be rendered to resemble ivory, tortoiseshell, marble, ebony, and semiprecious stones. Interestingly, Hyatt's company boasted in one pamphlet, "It will no longer be necessary to ransack the earth in pursuit of substances which are constantly growing scarcer."[7]

Then in 1907 Leo H. Baekeland, called "the Father of the Plastics Industry," developed *Bakelite*, the world's first synthetic, durable plastic.[8] Solid and sturdy, it was a favored material for high-value products like radios, telephones, toys, and game pieces; later it was used for wartime equipment such as pilots' goggles and some parts of firearms well into the 1940s. By then the improvements in chemical technologies that burgeoned in World War I were combined with the leaps in mass production made in WWII, setting the stage for the modern economy of plastics we see today.

Plastic Fantastic

It was this moment, when plastic went from being a prototype, premium material to a viable, cost-effective mode of producing consumer products, that manufacturers' uses for it became limitless. Injection-molding machines turned raw plastic powders or pellets into a molded, finished product in a one-shot process. A single machine equipped with a mold containing multiple cavities could pop out 10 fully formed products, like combs or flooring sheets, in less than a minute.

Coming out of wartime, output quotas long dedicated to government and the military machine were suddenly freed up for a plastics industry poised to break into an untapped market: civilians. After the war, according to one executive, "virtually nothing was made of plastic and anything could be." Soon synthetics factories were churning out Tupperware, Formica tables, polyester *fast fashion*, lifesaving Kevlar vests, and new toys like hula hoops, Legos, and Barbie. By 1960 plastics had surpassed aluminum, becoming one of the largest industries in the United States; in 1969 Neil Armstrong planted a nylon flag on the moon.[9]

Plastic has been the key enabler for sectors as diverse as packaging, construction, transportation, health care, and electronics. It's a simple way to mass-produce goods that once needed to be carved, welded, or blown out of heavier, more laborious material. Plastic packing and packaging material allows delicate items like food, medicine, and clothing farther distribution and easier handling. Polymers give body to common household items like insulation, piping, putty, and paint. Plastic increases consumer access to products and services

1.1 This table of main resin types, categorized by numbers 1 through 7, illustrates the pervasiveness of plastic in the modern economy.

both literally and financially, driving consumption. Innovation thrives with it, as industry has come to depend on it. **SEE 1.1**

The challenges of decoupling plastic from production and innovation as we know it would be eclipsed only by consumers trying to live without it. Today nearly everyone comes into contact with plastics, especially *plastic packaging*—its largest application, representing 26 percent of the total volume of plastics used.[10]

Plastic and the Linear Economy

While delivering many benefits, the current plastics economy has drawbacks that are becoming more apparent by the day. For instance, with more than 280 million metric tons of new, *virgin* plastic produced globally per year,[11] only 14 percent of all plastic packaging is collected for recycling. When additional value losses in sorting and reprocessing are factored in, only 5 percent of the material value of what we often use only once— *single-use plastics*—is retained for the next time around.

Plastic recycling has not kept pace with the continued demand for plastic production, which would be offset by the capture of more of this discarded material. And the problem is growing: today we produce 20 times more plastic than we did in 1964, and that volume is expected to double again in the next 20 years—and almost quadruple by 2050, the same year that plastics will outweigh fish in the world's oceans.

Nearly every product and packaging innovation has been brought into modernity with materials and designs that global recycling systems cannot handle, and consumer products companies are producing more materials that end up in landfills than ever before. Circular systems of *reuse*—vesting products with value and striving to keep them at high utility—have fallen in favor of largely linear ones that, despite the sophisticated science and technology behind them, view products and packaging as *disposable,* or designed to be thrown away.

Simply Circular

It wasn't always like this. Products and packaging used to cycle through a more regenerative *circular economy,* where, as in nature, things didn't go to waste. In contrast to the *linear economy,* this make-use-recycle-remanufacture concept creates value at each stage of a product's circular life cycle as recovered materials are returned to productive use. Here it is important to remember that waste in itself is a relatively modern idea that came about when it became more economically viable to produce new materials than to repurpose existing ones—and to burn and bury the rest. Up until the 1940s, when mass production, shifts in consumerism, and plastics came into play, things were actually quite circular. **SEE 1.2**

Dairy distributors provided reusable glass bottles that customers could empty and then leave on their doorsteps in the cultural motif we know as "the milkman." These glass bottles, which flowed through a system in which the producer was responsible for them (and owned them as an asset), had a high rate of reuse. So did durable containers provided by consumers for producers to fill with their purchases of other consumables such as oil, eggs, and cream.

What consumers didn't have delivered, they would shop for, and buying groceries and household items worked in the same way: either the patron or the producer provided reusable containers and wraps that could be returned, cleaned, and used

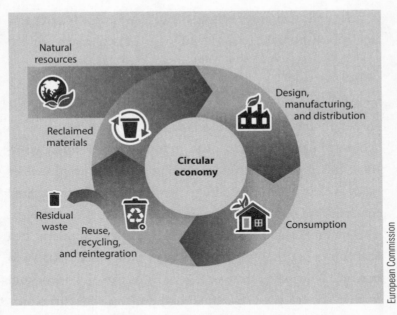

1.2 It was once intuitive that materials, resources, and products should reused, repaired, and recycled in what we call the circular economy.

again for the next batch. The concept of ***dual-use packaging,*** structurally designed to serve a function after first product use, got products off shelves by giving consumers "more for their money"; examples are condiments sold in decorative crocks and dry goods sold in reusable cloth bags or tin canisters. An emerging consumer culture encouraged more buying and selling, so producers sold "two products for the price of one," innovations of marketing and design that carried well through the Great Depression and far into the 1960s.

But companies eventually realized that sales need not be contingent on the flow of reusable containers. So, to make products easier to buy, use, and be needed again, they adopted packaging that took what couldn't be placed unwrapped in a cart, basket, or pack and made it marketable and easier and cheaper to buy.

Single-Use Packaging

Lighter and more portable than thick refillable glass bottles, commercial metal cans and tins used to store and preserve food (a revolution that afforded larger mobilization of armies, longer product lives, and massive reductions of the burden on supply chains) entered into production in the 1800s. Introduction of the beer can in 1935 got the ball rolling in terms of a viable way to mass-package and distribute beverages. The invention of the pull-tab in 1959 revolutionized the metal can as a convenient, lightweight vessel for beverages, with high function and recyclability (when recyclability was more of a priority, of course).

In 1915 the concept of carton-based packaging offered a lighter, paper-centric alternative to glass and metal. The patent on the first "paper bottle," called the "Pure-Pak,"[12] featured a folding paper box for holding milk that could be glued and sealed at a dairy farm for distribution. While the paper of traditional gable-top cartons could be reclaimed, today's carton technologies feature various combinations of plastic, metal, and paper; moisture barriers; and rigid plastic closures and fitments for function and convenience that are generally not recyclable. **SEE 1.3**

It's these highly affordable, single-purpose, self-service models of packaging that began to change inherently circular systems into linear ones. Over time people's expectations and

1.3 The milk container as evolved from delivery in reusable glass bottles (distributed from large aluminum cans), to the single-use paper carton, the plastic bottle, and the shelf-stable, multi-compositional, and largely unrecyclable aseptic cartons and pouches of today.

habits were shifting from reusing containers to throwing them in the garbage. Extending product shelf life and making it easy for people to simply buy milk in its own container whenever they wanted, these configurations would give way to mass production and distribution, as well as the expansion of the general store to the supermarket to finally the quarter-million-square-foot big-box stores we enjoy today.

Then, of course, came plastic. The use of plastic to package foods and beverages went from being an expensive technology to an affordable, economically viable practice when *high-density polyethylene* (**HDPE**, or #2 plastics) was introduced; in the wake of World War II, plastic production in the United States increased by 300 percent.[13] Compared with glass bottles, plastic's lightweight nature, relatively low production and transportation costs, and resistance to breakage made it popular with manufacturers and customers.

Durable, Long-Lasting—and Disposable?

Today the food-and-beverage industry has almost completely replaced glass bottles with plastic ones. In 2016 almost half a trillion **PET** bottles were produced, up from about 300 billion a decade ago, with continued growth in the forecast.[14] The demand, equivalent to about 20,000 bottles being bought every second, will increase to nearly 600 billion by 2021, according to the most up-to-date estimates from Euromonitor International's global packaging trends report.[15] Despite these projections, most of the plastic bottles produced today end up in the garbage.

Four categories of plastic packaging are tracked for linear disposal:

- **Small-format packaging** includes **sachets, tear-offs** (the thin plastic films on food containers), lids, straws, candy wrappers, and small pots (such as cosmetics containers) that tend to escape collection and sorting systems and have no economic reuse or recycling pathway. Small-format packaging represents about 10 percent of the market by weight and 35 percent to 50 percent by number of items.

- **Multimaterial hybrid packaging** currently cannot be economically recycled; these include stand-up food **pouches** and **aseptic cartons**. By combining the properties of materials, multimaterial packaging can often offer enhanced performance versus its monomaterial alternatives, such as providing oxygen and moisture barriers at reduced weight and costs. Multimaterial hybrids represent about 13 percent of the market by weight.

■ *Uncommon plastic* packaging materials, while often technically recyclable, are not economically viable to sort and recycle because their small volume prevents effective economies of scale; these include PVC, *polystyrene* (*PS*, or #6 plastics), and *expanded polystyrene* (*EPS*, also known as *Styrofoam*). Uncommon plastics represent about 10 percent of the market by weight.

■ *Nutrient-contaminated materials*, from dining disposables to coffee capsules, are often difficult to sort and clean for high-quality recycling. This segment includes applications and configurations that are prone to be mixed with organic contents during or after use.

Instead of milk bottles, we now have milk bags. Instead of getting our soda in refillable glass bottles or recyclable aluminum cans, we buy a plastic bottle to take with us to eventually toss. Every step on this progression has brought with it less and less recyclable packaging; the recyclability of each of these packages is effectively halved with every step, with all *flexible packaging* being completely nonrecyclable.

The Role of Business in the Circular Economy

Industries and businesses (and, in part, the consumers who demand it and the governments that allow it) have driven the shift away from the naturally circular patterns of yesteryear.

Thus companies and major brands are the ones in a position to compel the change forward toward more regenerative business practices. Although the global recycling infrastructure is inefficient and the world economy continues to view material as useless after one use, there is opportunity to capitalize on these gaps in the same way that plastic revolutionized production and continues to drive consumption in the first place.

Brands, governments, celebrities, and *nongovernmental organizations* (**NGOs**) are today promoting a host of innovations that provide real solutions to the plastics problem. **Reusable** packaging is part of the answer, but designing a plastics reclamation system that works and reinventing the types of plastic packaging that make technologies possible but almost never get captured are objects of our search. Recent New Plastics Economy Innovation Prize winners included a compostable multilayer material from agricultural and forestry by-products (perfect for food packaging) and a magnetic additive that can be used in lieu of foils in moisture barrier technologies in aseptic cartons and pouches for a fully recyclable package.[16]

Calling for an end to the current single-use plastic–reliant product economy—despite mounting knowledge of the environmental and social justice issues we know plastics to present—is much, much easier said than done. So instead of simply setting out to change what we produce and consume through design, we must strive to also change the way we participate in the product economy.

There are endless opportunities to establish circular systems where they currently do not exist and to strengthen them where the groundwork has been laid. These are the practices that will differentiate you from your competitors and prepare you for resilience and growth in an uncertain future.

CHAPTER 2

Where Did Public Recycling Come From, and Where Is It Going?

Christine "Christie" Todd Whitman

President, The Whitman Strategy Group; Former
Governor of New Jersey; Former Administrator,
Environmental Protection Agency

TAKING A LOOK AT THE HISTORY OF PUBLIC WASTE MAN-
agement systems can provide some context for what is
missing from them today and how we might improve them.
Before any formal collection and recycling programs, it was
intuitive that old and discarded materials would be reused.
Materials were perceived as having great value, and new—or
virgin—materials were not only costly but difficult to extract
and refine. Just imagine having to harvest more copper from a
mine when you could just melt down scraps from old or broken
copper products you already had handy.

As technologies spread across civilizations, craftsmen
and artisans developed trades and demanded natural resources
in greater and greater quantities. For example, after the first
examples of paper as we know it today (post-papyrus) were
recorded two millennia ago in ancient China,[1] methods for
recycling it soon followed. The more people used paper to
record drawings, writings, and history, the more of it they

needed. To meet demand, production and recycling were one and the same.

This was the way for most of history, and it remained so as papermaking spread from China to Japan to North Africa and Europe. In the United States in 1690 (still British America), the Rittenhouse Mill near Philadelphia collected used parchment and worn-out cotton and linen rags, each relatively expensive and limited in supply, and converted them into handmade paper. It was intuitive that products would be reused. Recycling and reuse weren't laws but rather rules of life. It simply made sense.

Necessity Is the Mother of Invention

Everything from dead animals to soup bones to scrap metal could supply new production. *Waste pickers*—or rag-pickers, as they were called, along with wharf rats, tinkers, rag and bone men, and mudlarks[2]—salvaged materials found in the street and in waste bins and sold them to wholesalers, who sold them to mills and artisans. Industrialization and urbanization facilitated these informal recycling economies,[3] a tradition today represented by "canners" and waste pickers in developed and developing countries. **SEE 2.1**

While the growth of cities and industry is a central theme to the increasing demand for material, nothing places strain on economic and material resources quite like war. Less than a century after the Rittenhouse Mill went up, the American War of Independence suddenly called for more-systematic methods of collecting items for reuse and recycling into new products. Not only did trade between the United States and

Eugène Atget/Wikimedia Commons

2.1 A rag-picker in Paris, ca. 1899–1901.

Britain cease but the former, much younger country did not have the material-sourcing infrastructure in place to sustain the livelihood of its domestic population, let alone to battle and break away from the British.

So, necessity became the mother of invention: the invention of public (albeit not yet mandated) recycling systems made gathering materials the responsibility of citizens—and one they carried out gladly. People reused and collected not just for their own immediate needs but for the needs of a greater movement. Citizens voluntarily participated in the collection and processing of scrap metal, paper, cloth, and other used items, incentivized by the patriotic nature of these activities and their contribution to the war effort. Iron kettles and pots, among other things, were melted down for armaments. Recycling and reuse wasn't a waste management activity but rather a method to source material, of which there was a profoundly limited supply.

Fast-forward and compare this with World War II, when Americans were famously encouraged to collect scrap metal, paper, and even cooking waste,[4] presumably for use in the production of soaps and, more excitingly (and ominously), explosives. Through the Salvage for Victory campaign, cities were given recycling quotas to fill (some dedicating their sanitation departments to pick up materials weekly),[5] so families were called on to recycle tin cans, collect scrap paper door to door, and ration their consumption of goods on the home front. Some experts now say that the biggest contribution of these austerity measures and material collection drives was boosting morale, as many of these items were not used, but this period planted the seeds of public recycling programs in the United States as we know them today. **SEE 2.2**

Across the pond, our allies in Great Britain also focused compulsory government-backed programs on the recycling of materials as part of their National Salvage Campaign. Launched in 1939 at the very outbreak of the war, what was initially a

Everett Historical/Shutterstock

2.2 Americans were encouraged to collect scrap metal, paper, and even cooking waste for use in arms production during World War II. Here a housewife of the era sifts bacon grease to be processed into ammunition for the war effort.

campaign only strongly encouraging household collections became a strictly enforced and controlled operation in which people who refused to sort their waste were fined and faced time in prison. Newspaper advertisements explained how every ton of paper saved was equal to 2,956,800 cigarettes, 12,000 square feet (1,100 m²) of ceiling board, 17,000 sheets of brown wrapping paper, or 201,600 books of matches.[6]

Before the war less than 900 metric tons of scrap paper was salvaged each week in Britain, rising to nearly 400,000 metric tons in 1942. Sixty percent of all new paper derived from recycled sources.

Consumerism in Peacetime

When the war ended with the Allied victory in 1945, aspects of the UK salvage scheme were kept to help transition to peace and later help with an economic downturn. But even with these financial problems toward the end of that decade, it became clear that the social practice of salvage at that point was very much tied to the campaign for national survival.[7] Values change in the transition from war to peace. Without the pressure of patriotism, Britain, the United States, and the rest of the world slipped into a throwaway culture and emerging modes of fashion, technology, and consumerism.

Once-rationed products like shampoo, cigarettes, food, and clothing were now available, in tubes, pre-rolled, packaged, or hanging in every store window you passed on the main drag. In a postwar world of plenty, disposable goods like single-use dining wear and packaged snacks, cosmetics, and razors became symbols not just of affluence but also of abstract ideas like freedom and hygiene.[8] Consumers took to disposable items that would allow them to buy again and again and again, and they viewed their ability to waste as a comfort in peacetime.

Consumption became the center of the American Dream[9]—producers were ready to sell it, and town and city governments were ready for the revenue. Production technologies and the development of new products, new modes of shopping, and new advertising techniques assaulted any culture of recycling or ***bricolage*** (the improvisation of items from secondhand materials) that kept reuse, repair, and reclamation in favor. Returning to their prewar exploitation of virgin resources, manufacturers and brands churned out plastic products and packaged goods that, in a booming economy, were not regarded as items that needed to be collected and repurposed. Again, producers wanted to encourage consumption, and without government regulation or sanctions they were able to do so quite freely.

Public Recycling and Waste Management

By the time public recycling began popping up in the 1960s and 1970s, it was to deal with the massive amounts of waste produced as a result of this overconsumption. Landfills that were supposed to have a life span of two decades[10] remained active longer due to relentless garbage production and a lack of alternatives. Recycling became a waste reduction effort in a seemingly uphill battle for the environment. Garbage was now doing more than overflowing from trash cans and landfills; it was littering oceans, national parks, and public areas. The public was becoming aware of the toxicity and wastefulness of synthetics.

One turning point in the history of recycling in America can be attributed to the famous public service announcement by Keep America Beautiful, a national community improvement nonprofit organization: "People start pollution. People can stop it." Shortly following the celebration of the first Earth Day in 1970,[11] the ad framed recycling and waste as a problem

GET INVOLVED NOW. POLLUTION HURTS ALL OF US.

You can help by becoming a community volunteer. Write:
Keep America Beautiful, Inc.
99 Park Avenue, New York, New York 10016
A Public Service of Transit Advertising & The Advertising Council

Ad Council

People start pollution.
People can stop it.

Keep America Beautiful

2.3 The famous 1971 "Crying Indian" public service announcement by the Keep America Beautiful coalition ushered in a turning point in the history of recycling.

for consumers, rather than one for the companies that manufactured the items.[12] This was reflected in the structures of public recycling programs. Separate *waste streams* for metals and paper required consumers to separate their garbage, mobilizing labor for no pay and investing in infrastructure through the taxes that citizens put into the system. **SEE 2.3**

Waste was the catalyst for the formation of the US Environmental Protection Agency (EPA) in 1970 and passage of the Resource Recovery Act by Congress in 1976. The first *materials recovery facilities* (*MRFs*, pronounced "murfs") increased the ability of municipal agencies to collect and process recyclables at volume, and curbside recycling programs offered consumers the ability to recycle from the comfort of home, as opposed to driving bottles and cans out to the nearest materials processor.

Environmentalism and recycling became mainstream in the same way plastic, consumption, and waste did: through advertising, marketing, and public relations campaigns.

Schoolchildren and young people learned about it in school, and the media brought it into the home through television, news stories, and on-pack messaging. The three-arrow recycling symbol came into being through a submission from a college student for a design competition.[13] Recycling was becoming part of public consciousness, although a majority of the population (more than 90 percent)[14] still did not participate in public recycling due to a lack of personal benefit, poor enforcement of laws, and the out-of-sight, out-of-mind nature of linear trash systems.

The introduction of **bottle bills**, also known as *container deposit laws*, provided a monetary incentive to return beverage containers for recycling, with some success. In the first part of the twentieth century, consumers saw getting their money back as a normal part of purchasing something that came in its own bottle,[15] but the first US state to pass a beverage container deposit law was Oregon, in 1971. Today the 10 US states with bottle bills boast a 70 percent average recycling rate, compared with an overall rate in the United States of 33 percent. The challenge is that bottle bills not only are not growing but are actually declining due to industry pressure, with Delaware and Missouri recently repealing theirs.

The next decade brought on what came to be known as the "landfill crisis of the 1980s,"[16] as a throwaway culture, major manufacturing of synthetic items (picture shiny exercise apparel, VHS tapes, and video game consoles), and dependence on linear disposal brought landfill use, which was inexpensive at the time, to a fever pitch. By the end of the eighties, landfills were either closing or massively increasing their rates.

Taking Responsibility

Interestingly, this is also about the time that companies and brands were realizing the business potential of marketing to environmentally minded people through what would become known as *corporate social responsibility*.[17] Manufacturers monitoring working conditions and combating child labor, as well as the first fair-trade labeling initiatives, gave consumers the opportunity to express their values and "vote" for their social preferences through what they purchased.[18] Yet consumers still did not expect business to take responsibility for the end of life of the packaging and products flowing through these markets. Companies and brands continued their business as usual, despite consumer interest in recycling.

It wasn't until 1990 that the US recovery rate for recycling would surpass 15 percent. The 66-page Environmental Protection Act of 1990 outlined new guidelines for the disposal of waste, and cities across the country were implementing their own recycling programs. As they waited for funding to get to their regions, some enterprising teachers and principals in school districts started their own programs. One district manager was quoted as saying, "I did not start recycling until my children showed me what it meant to recycle."[19] Public recycling programs were embraced by residents in a way that critics would not have anticipated. **SEE 2.4**

As the population and interest in recycling grew, so did the time, money, and resources required to collect the increasing amount of material that could be recycled. The introduction of *single-stream recycling* in 1995, wherein municipal

Seattle Municipal Archives/Wikimedia Commons

2.4 Recycling really took off in the 1990s, when it became a mainstream activity in which people actively participated.

solid waste is collected in a single bin, rather than separated by material type, increased consumer participation because it was easy to understand. It also saved cities money upfront, as it's easier to dump one can of waste into a collection truck with a single compartment.

Single-stream recycling, however, required added separation on the back end, so it diminished the quality of recyclables. Glass, for example, dumped into a big bin, onto a tipping floor, and then on conveyor belts at MRFs, would often break; not only would this damaged material not get recycled but it would contaminate bales of paper. Not requiring consumers to separate their waste also increased the amount of nonrecyclables placed in the bin, further adding to the labor required to not only separate but also dispose of unrecyclable waste. A practice that carried through to the present day, single-stream recycling still has many of these issues.

Recycling in the United States: Stop and Go

Recycling went viral in the 1990s, but instead of succeeding, it died under its own weight. A strong market demand for used paper and other recyclables eventually bottomed out due to oversupply,[20] causing a booming market to collapse.[21] There was a glut of material with not enough waste management infrastructure, energy, or resources to maintain successful models for participation, like single-stream recycling, while ensuring the quality of the resulting *feedstocks*. Plans to increase recycling research, public education, and enforcement hit funding obstacles, which caused municipalities to dial back their efforts. From 1990 to 1997, plastic packaging grew five times faster by weight than plastic recovered for recycling.[22]

The EPA reports that between 1980 and 2010, the national recycling rate grew 0.8 percent per year on average. In comparison, between 2010 and 2014 growth had drastically fallen to 0.1 percent per year.[23]

Recycling rates would plateau at 34.5 percent, and even this can be attributed to population growth. The price volatility of oil in the early 2000s did not compel US consumers to recycle more plastic (though it did cause a good deal of stress at the gas pump). In fact, the Great Recession of the latter part of the decade, felt particularly in North America and Europe, saw *increased* consumption of cheaper *consumer packaged*

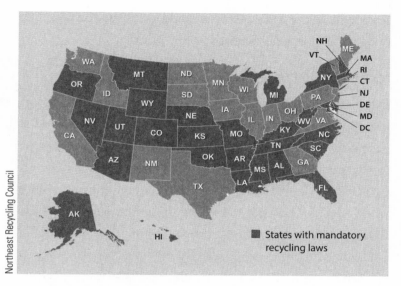

2.5 The Northeast Recycling Council's map of states with mandatory recycling in the United States (2017) is a window into the country's fragmented laws on recycling, which is a low priority for governments state to state.

goods (*CPG*)—such as prepackaged foods and smaller units of household consumables like detergent and dish soap—in increasingly unrecyclable packaging. **SEE 2.5**

Today in North America, recycling is seen as the right thing to do, but it no longer functions as a means to supply industry with material, so it makes little business sense. In the United States, there is such a fragmented view on waste disposal that mandatory, enforceable recycling laws vary greatly from state to state, even town to town. At this point recycling won't dent the islands of garbage the size of Alaska now floating in our oceans.[24] Governments face pressure from industry to rescind recycling goals and polices and to cut back on recycling budgets, as this area is viewed as low priority.

Successful Models Do Exist

Every region, government, and populace has different priorities, but the key to moving forward is to look back at what used to work and see how we might apply those concepts today and in the future. During a stay in Scottsdale, Arizona, I got a strongly worded note from the city's solid waste office when I accidentally put the wrong thing in the recycling bin. It reminded me of how not properly sorting your waste could land you a night in jail or a hefty fine back when recycling was a very real, viable, life-or-death material-sourcing system during World War II.

In Brussels, where my daughter lived for several years, public works personnel went through the color-coded trash bags before they would accept them for pickup. Unlike in the United States, where municipal pickup personnel work to get the waste to the facility as quickly as possible, in Belgium it's possible that your trash won't be taken away at all if you don't separate it, providing an incentive to sort and make recycling worth it for the city. The onus is placed first on consumers to separate their waste and then on public servants to enforce it. The extra labor and logistics requires funding, which the city subsidizes through tax dollars.

In New Jersey, where I served as governor, laws and law enforcement have been integral to the effort for more-sustainable waste management systems built on transparency and accountability. Woodbury, New Jersey, was the first city in the United

States to mandate recycling, with the first municipal curbside garbage truck pulling a specially designed trailer. Met with some initial pushback from residents (people reportedly threw trash on the mayor's lawn in protest at first), within three months the program reached 85 percent compliance and became a national, replicable model for curbside recycling. This, like all new initiatives, had an adjustment period.

The Garden State has also pioneered the passage of laws requiring the cleanup of solid waste processing facilities along rail lines, closing a federal loophole that prohibited states from enforcing environmental, health, and safety regulations at such sites.[25] This is significant in that the state and local governments were able to step in where federal regulations would not, to manage waste in the interest of resident populations. Situations like this beg the question of how recycling and waste management initiatives may best be enforced at the local, state, and federal levels and where leadership is most effective.

With private companies like TerraCycle, we are now getting into the recycling of what had previously been considered unrecyclable—like ocean plastic transformed into usable, marketable products that people can buy and recycle again. Looping the activity of recycling back to its origins as a material-sourcing system reinforces its value and capitalizes on its full range of environmental and industrial benefits. While the public sector still struggles to keep up with the growing rate of consumption, often battling with industry for the ability to regulate, private entities can act freely to improve recycling. **SEE 2.6**

2.6 With private companies like TerraCycle, we are now getting into the recycling of what had been previously considered unrecyclable, such as empty pens, markers, and other writing instruments.

Business and government are in a position to work together to motivate individuals to participate in recycling systems, both private and public. Recycling capabilities vary from region to region, but we must think holistically about the need for better waste and resource management. The state of the recycling industry around the globe is fragmented, as are the specific cultures and needs of each region, but the world's problems with waste are the same.

CHAPTER 3

The State of the Recycling Industry

Jean-Marc Boursier

Group Senior Executive Vice President and
Chief Financial Officer, Finance and Recycling
& Recovery (Northern Europe), SUEZ

THE GLOBAL RECYCLING INFRASTRUCTURE IS FRAGMENTED. There is no clear, across-the-board guidance or top-down leadership about what can be recycled and how to solve for waste. At the national, state, provincial, and local levels in sovereign bodies around the world, there are different regulations and public policies ordering the behavior of businesses and consumers. Because of this, recycling is often viewed as voluntary, and people who do *recycle*—or repurpose a waste object by valuing only the material from which it is made—may be confused about or unaware of what is accepted.

Day by day what can be recycled locally will change. Recently, China (which previously purchased approximately two-thirds of North America's recyclables) passed legislation called the *Green Sword,* which banned the import of 24 categories of solid waste, including certain types of plastics, paper, and textiles, effectively gutting recycling markets worldwide

3.1 What can be recycled changes day to day. China, which previously bought approximately two-thirds of North America's recyclables, recently passed regulations on foreign garbage.

and rendering even more packages and products unrecyclable to local citizens.[1] **SEE 3.1**

Waste by the Numbers

Billions of dollars of valuable, usable materials are lost because of the way global recycling systems currently operate. Worldwide municipal waste production is set to triple in volume from 3.6 billion metric tons today to 11 billion metric tons in 2100, less than a century.[2] In France, where SUEZ, the largest waste management company in Europe (similar to Waste Management, Inc., in the United States) is based, municipal waste volumes doubled in the 40 years between 1960 and 2000, from 175 to 350 kilograms per resident per year.[3] At the same time, plastics production in Europe has grown by 50 times in 50 years, going from 1 million tons in 1950 to 50 million tons today.

3.2 The quantity of sorted materials, such as these post-consumer plastic bottles, is much higher than the capacity of industry to incorporate the recycled raw materials in production.

There remains a decreasing economic incentive to recycle, as front-end costs (collection and processing) exceed the sale price of materials on the back end. The European and North American markets are totally unbalanced, for instance, in that the quantity of sorted materials (i.e., plastics, paper and cardboard, and ferrous and nonferrous metal) is much higher than the capacity of industry to incorporate those **recycled raw materials (RRM)** into their production lines. Further, most materials won't even get to that point of sortation, as it currently costs more to collect and process RRM (also known as **post-consumer recycled,** or **PCR**) materials than it does to simply send them straight to landfill or incinerator. **SEE 3.2**

Changing the vision around packaging waste to make it work for us as a resource requires a hard look at the varied nature of global waste management programs. As it stands, the world's diversity of government structures, sustainable development priorities, and demand for recycled materials creates a challenging and fragmented matrix of regional programs. Private industry, public policy, and consumer activities all play their roles in this.

Who Can Improve the Recycling Infrastructure?

Of each of these players, private industry is in the best position to change the state of the recycling industry in the shortest period of time with the most impact. Not only does industry's support for better recycling mitigate the risk of the political process of governments (which often face pressure from businesses to make it easier for them to operate and profit as they please) but private companies are the ones putting out the products for consumers to demand.

Businesses face considerable challenges to developing more environmentally friendly packaging options and investing in systems that make more of their packaging recyclable. In the current global infrastructure, this is plain. Especially in lower-income communities and economies in the developing world, pouches and sachet packaging, for example, make it possible for products to sell. Investing in packaging by using heavier, more recyclable materials, or even alternative materials like compostable bioplastics, drives up costs.

If private industry has to work closely along the matrix, the public sector needs to enact policies that favor the use of RRM through the intensifying of green procurement and legislation that promotes (or even forces) the use of recycled content. Fiscal incentives to boost demand should also be considered, such as lower value-added tax rates on recycled polymers or a carbon tax. Then cities have to apply and enforce these policies.

Citizens also have a decisive role in the recycling industry as it relates to the circular economy: they have to be vigilant when they buy products and rigorous with themselves when it comes to sorting and diverting valuable material from the landfill.

Recycling Is Hindered by Fragmentation

Most countries do not have a national program mandating participation in recycling; in developing countries this is often the case, as waste treatment is not a priority for development. This lack of access often leaves consumers with few choices other than landfilling their waste or seeing it pile up as litter. Many postindustrial countries and world economic powers, however, are in the same boat.

The United States has no federal or national programs mandating recycling. Americans, despite being only 4.4 percent of the world's population, produce 20 percent of the world's garbage[4]—on average 2.6 kilograms of garbage each day.[5] ***Packaging and printed paper (PPP)***, which represents about 41 percent of the 240 million metric tons of municipal solid waste generated each year, is the largest contributor to the growing heap.[6] Some 96 million metric tons of PPP—steel, aluminum, glass, plastic, wood pallets, paper, and paperboard—are discarded each year. About half of this amount is recycled, while the other half—about 47 million metric tons—is disposed of;[7] 75 percent of this waste goes to landfills, and 25 percent[8] goes to one of 77 US facilities that burn garbage and possibly produce energy.[9]

The Resource Conservation and Recovery Act of 1976 includes language that gives the Environmental Protection Agency the authority to control and define hazardous waste, and it details a framework for the management of non-hazardous solid wastes (municipal solid waste and industrial).[10]

There is no top-down, federal enforcement compelling consumers to recycle, however, and there are no incentives for industry to make a more recyclable product or package. Public and municipal recycling programs are left to the states, most of which are delgated to counties and towns and cities, across which curbside recycling schemes differ. For example, one city may accept PVC, or #3 plastics, and the town over, despite being in the same state and only a few miles away, may not. Thus a package that is recyclable in your city may not be recyclable in the next.

Lack of Enforcement

With the exception of a few US cities, there is no penalty for separating waste incorrectly, and the ones that have imposed fines have been met with pushback. For example, the City of Seattle attempted to enforce a ban on food waste and recyclables in trash bins by having collectors take a look and flag noncompliant bags with a red tag and ticket saying to expect a $1 fine on the next garbage bill. Residents called it an invasion of privacy, and after a lawsuit this aspect of the mandatory composting and recycling scheme was blocked. **SEE 3.3**

Cut fees (if possible)   Pay per bag   Incentivize what is right

WasteZero

3.3 Some American cities have implemented monetary incentive structures to encourage people to recycle through positive reinforcement.

Conversely, some US cities have taken a positive approach by implementing monetary incentives that encourage people to recycle. San Francisco, called "the Silicon Valley of Recycling,"[11] also prohibits food waste in trash, employing a three-color bin system with a "pay as you throw" framework that offers residents economic subsidies for sending less waste to landfill.[12] Few things get people to understand concepts as the norm as well as saving money. This extends to business, which has access to a range of services and assistance in implementing business recycling and composting programs from San Francisco's environmental protection agency, SF Environment.

Not all towns are as progressive as San Francisco. Some areas in the United States don't have curbside recycling programs at all, putting the onus on residents to find a solution in surrounding towns or the private sector. In these areas, in the United States and otherwise, recyclables and nonrecyclables tend to go straight to landfills or unregulated dumps.

Similarly, Canada too has no national program mandating recycling. What it does have is Environment and Climate Change Canada (ECCC), the federal department that operates in part to coordinate recycling at the provincial level.[13] Though ECCC shares responsibility for waste management in Canada, each of its 10 provinces and three territories administers recycling programs, which then delegate responsibility to the individual municipalities.

Very few cities and towns in Canada impose penalties for putting the wrong items in the bin, but with work this will change. Toronto, the country's largest city, finds 25 percent of what's collected from recycling bins to be contaminants, which costs the city money to sort out and deal with.[14] Like in the United States, each municipality has different ways of enforcing recycling and accepts different waste streams for recycling. The nation's last recorded recovery rate is 27 percent.[15]

Waste: An Issue of Global Equity

Mexico has no formal national recycling program or infrastructure and no proper waste management in its slums. Improvements are being made in terms of a growing number of recycling plants for PET plastic;[16] however, the waste industry in Mexico is currently controlled by the richest 1 percent of the country's population. There is a $24 billion opportunity in the recycling industry in Mexico, and there are not enough opportunities for low-income communities.[17] In a country where the average individual generates 1 kilogram of waste per day (or 90,000 metric tons total as a country per day), the overall recycling rate is only 3.3 percent.[18]

Recycling doesn't just need major investment of time and money to improve logistics, residential access, technology, and enforcement systems; it needs a reallocation of controls to make it more equitable as an industry. A major source of revenue and real jobs creation, recycling is hindered by monopolies all around the world, preventing the systems from moving forward. Governments can step in here to regulate, and businesses can band together to compel change through industry coalitions and trade unions.

In general, the amount a nation spends on environmental protection tends to be positively correlated to national **gross domestic product (GDP)**. GDP is the sum of consumption, investment, government spending, and net exports, and it's an indicator of the economic resources devoted by taxpayers to environmental protection. Business, government, and household sectors[19] strained for budgetary assets are less equipped to invest in recycling, which is often low on the list of priorities next to, say, water infrastructure, electricity, and transportation.

But the agendas of developing-world economies do include environmental protection and sustainable materials recovery. Composting and material sorting are focus areas in Ecuador, which I visited a few years ago with the SUEZ Foundation. Its central Chimborazo Region currently dumps more than 80 percent of its municipal solid waste into the environment, disposed of untreated in open, unregulated landfills. Pilot initiatives to develop organic fertilizers from food waste (generated in the amount of 5.5 metric tons daily) aim to manage the trash influx and generate income.[20]

In Meknès, one of four imperial cities in Morocco, the long-unregulated landfill facility that accepted 165,000 metric tons of waste annually is in rehabilitation through a 20-year contract with SUEZ,[21] which is working to design, build, and operate a waste elimination and recycling facility for materials recovery.[22] In addition, former waste pickers became proper employees of our group, which made me particularly proud.

Moving toward a Circular Economy

Implementing waste treatment policies in the context of money spent up front is always more expensive than doing nothing. Investment costs are rising as changing waste streams demand more-complex sorting systems and more nonrecyclables come through processing facilities, contaminating streams. Collection receptacles, transportation vehicles, logistics personnel, educational communications, and materials recovery and processing equipment are only as viable as the resources behind them. These investments are supported by three different avenues: waste treatment services, taxes and levies, and increased product prices as part of *extended producer responsibility (EPR)* policies. **SEE 3.4**

In Europe regulation and public policies are speeding awareness about waste production and forcing a move toward more recycling and recovery. The Circular Economy Package of the European Commission, the European Union's politically independent executive arm, includes legislation on waste with strong commitments on circular design by manufacturers and companies, strategies for handling plastics and chemicals, and funding for innovation.[23]

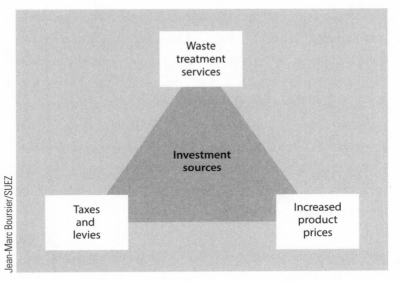

Jean-Marc Boursier/SUEZ

3.4 Investments in waste treatment are supported by three different avenues: waste treatment services, taxes and levies, and increased product prices as part of extended producer responsibility schemes.

A more unified front intended to improve the quality of recycled material ties back into creating stronger demand for RRM. In 2014 less than one-third of Europe's plastic waste was sorted, with a bit less than a third captured for energy recovery. The rest ended up in landfills. Only 7 percent of the 45 million metric tons of plastic used in production per year is recycled polymer; the other 93 percent is still virgin material coming from fossil fuels. The Circular Economy Package aims to get these up to 65 to 70 percent municipal waste recycled by 2030.

Across the European Union's 28 countries, the 2015 average recycling rate was 45 percent,[24] each nation employing its own recycling schemes to increasing recycling rates by improving framework conditions and providing economic instruments. With a countrywide recycling rate to match the EU average, some municipalities in the United Kingdom employ up to nine bins in their residential recycling programs,[25] which are meant to better sort waste streams at the source to reduce contamination. Criticisms of this setup, also employed in Germany, include resident confusion; but when you compare this with single-stream recycling, sorting at the source provides added protections for the quality of the resulting recycled raw materials.

The Nordic countries have some of the highest recycling rates in Europe. As in Great Britain, some municipalities boast a number of recycling bins or bags for separation, glass and plastic each getting up to three bins for their various colors and resin types (multi-compartment bins are also used). This is combined with voluntary and involuntary EPR schemes in which manufacturers and brands are accountable, or pay, for the recycling of their products and packaging.[26] In Sweden special trucks go around cities and pick up electronics and hazardous waste like chemicals, and residents take their larger waste, such as broken TVs and furniture, to recycling centers on the outskirts of the cities.[27] **SEE 3.5**

The key to the success of the Nordic countries' recycling platform is the fact that they tend to be consistent across the country, with the same types of bins for all, including households and businesses. Other factors include high separation at collection, consistent signage, consistent bin colors, stickers identifying wrong materials, fines for non-compliance, and accessibility of bins and drop-off points. The rate at which

Szary Wilk/Wikimedia Commons

3.5 The Nordic countries have some of the highest recycling rates in Europe. As in Great Britain, some municipalities boast a number of recycling bins or bags for separation.

recyclables can be exchanged for money compared with other countries is also quite high,[28] so most discarded items are viewed as valuable.

Bending the currently linear line to close the loop will also depend on how products are labeled and designed. This requires better cooperation among brand owners, manufacturers, and the waste management sector. The United States has the private How2Recycle label, Germany has Der Grüne Punkt (the Green Dot), and many countries around the world use resin identification numbers on their plastic packaging, but governments and manufacturers could do more to educate and enforce these with the public, as many consumers are still confused about what these all mean. **SEE 3.6 AND 3.7**

3.6 The USA's
How2Recycle label.

3.7 Europe's Der Grüne
Punkt (The Green Dot)

In parallel to existing municipal waste collection systems, solutions do exist. Secondary metals, glass, and paper are currently price competitive in the market with virgin-like quality, but for plastics, even if the price is satisfactory, the quality of many recycled polymers in many cases doesn't come close. The development and support of a secondary plastics market is essential to increase the demand for RRM, in turn giving value to investments in infrastructure and logistics. But market forces alone have been insufficient in driving a comprehensive recycling solution. We need a long-term and ambitious policy framework to provide legal and financial certainty.

Governments already regulate aspects of business to ensure the safety of consumers. Why can this not be true of recyclability? The EU Circular Economy Package isn't perfect, and it will take time, but it at least starts on a path of circular thinking that consumers, businesses, and local governments can each replicate and apply. The transition to a circular economy can be achieved only when all actors work together to design win-win solutions.

Challenges lie ahead—but they also offer a lot of opportunities to transform the economy, create jobs, and protect our environment by reducing the use of fossil energy and carbon dioxide (CO_2) emissions. Taking responsibility for our packaging as consumers, businesspeople, and agents of government is the first step toward a comprehensive resource management system. And personally, this transformation toward a more sustainable future is what motivates me most.

Who Is Responsible for Recycling Packaging?

Scott Cassel
Founder and CEO, Product Stewardship Institute

WE AMERICANS OFTEN TOSS PACKAGING IN THE TRASH without much thought. As stated previously, even though we are only 4.4 percent of the world's population, we produce 20 percent of the world's garbage; much of it is ***packaging and printed paper* (PPP)**. Proportionally, that's a lot. **SEE 4.1**

Everyone who touches packaging has a role to play in ensuring that its value is captured and that it doesn't add to the world's pollution. But who should be first in line to take financial responsibility? Is it the producers who make it, the retailers who sell it, or the cities where all of this takes place? Or is it, perhaps, the consumers who choose to buy it?

Despite the global fragmentation of laws and waste management systems, government has a major role in changing consumer and industry behavior when it comes to wasteful packaging. We see that especially when encouraged through a mode we all understand: money—in the form of fines,

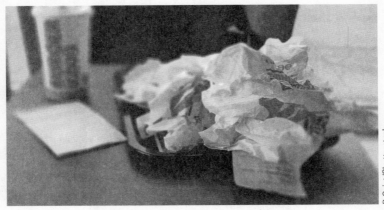

4.1 Although Americans constitute only 4.4 percent of the world's population, we produce 20 percent of the world's garbage.

penalties, and incentives. When such levers are put into place, people improve their behavior quickly and dramatically.

Businesses are subject to vast amounts of government regulation in the interest of protecting consumers and ensuring a level playing field. Among other things, laws today require that labels and packages provide more facts about the contents inside[1] and aim to preserve our health.[2] In the world of consumer packaged goods, we see this with certified-organic and organic-transitional labeling, specific ingredient bans, fair-trade sourcing conditions, and acceptable levels of certain chemicals in products and packaging.

But can you think of any laws regulating the end of life of the packaging itself? Many such laws exist around the world, especially in developed countries. In the United States, some mandatory recycling laws exist at the state and local levels, but federally there are none.

Challenges to Recycling Laws

Business brings tax revenue and jobs to cities, states, and countries, so business interests often drive government regulations. But there are regulations that businesses don't like, mainly those that cost money and reduce the ability to maximize profits. For most businesses and entrepreneurs, regulations are often viewed as financial and legal barriers to growth, and corporations see it as an obstruction to their desire to maximize return for their shareholders.

While their member companies finance recycling and resource management systems throughout the world, trade associations such as the American Institute for Packaging and the Environment and the Grocery Manufacturers Association have opposed legislation in the United States under the philosophy that packaging disposal, recycling, and litter cleanup costs should be the responsibility of government.[3]

Thus recycling laws get left to the states in the form of bottle bills; the banning of Styrofoam containers, plastic bags, and drinking straws; and guidelines for the disposal of *e-waste*, paint, and pharmaceuticals. This means the make-use-dispose linear economy pipeline currently employed around the world becomes only more and more pronounced and entrenched as time goes on. Year after year manufacturers create new products at a fraction of the cost of their predecessors, so more people now own more and more things—things that have a shorter and shorter useful life. **SEE 4.2**

Policies like ***bottle bills*** tend to get pushback from industry. Although bottle bills provide consistent, high-quality recycled material, industry often argues that such regulations

Take   Make   Dispose

Product Stewardship Institute

4.2 The take-make-dispose economy for packaging only grows more pronounced as businesses continue to make products that are unrecyclable—and that are a fraction of the cost.

are cumbersome, expensive, and a logistical nightmare. As a result, they end up not being passed; in the end governments can regulate only to the point that society is willing to bear.

Even with broad availability of recycling programs in much of the United States, the recycling rate for PPP—including traditional curbside recyclables such as aluminum, glass, plastic, paperboard, newspapers, phone books, and office paper—has been stagnant for the past decade.

Extended Producer Responsibility

One solution may be to shift the responsibility from taxpayers and governments to product manufacturers, as they have the distinct ability to choose what package forms they use for their products. With this in mind, should they be the primary responsible party to pay for the proper end-of-life management of their products and packages, even if this cost finds its way to the consumer in the end?

Extended producer responsibility (*EPR*) is the policy concept that extends a manufacturer's responsibility for reducing upstream product and packaging impacts to the downstream stage, when consumers are done with them. There are more than 110 EPR laws currently in place for over 13 product categories in more than 30 US states. The United States, however, is currently one of only three nations of the 35-member Organisation for Economic Co-operation and Development that does not have an EPR system specifically for packaging in place or under development.

EPR packaging laws have been in place for up to 30 years in 11 countries in Asia, South America, and Africa, as well as in Australia, 34 European nations, and five Canadian provinces. While not all EPR programs are alike, the best ones are not voluntary in nature and produce recycling rates far higher than what we have experienced in the United States. British Columbia and Belgium, both of which have EPR packaging laws in place, have attained nearly 80 percent PPP recovery. **SEE 4.3**

Voluntary industry-led programs, while laying a foundation for collection and recycling systems, rarely lead to systemic changes that significantly increase the quantity and value of the materials collected, and they do not provide a sustainable funding source across all producers in a certain category. For instance, although voluntary initiatives to collect plastic films at retail outlets have helped reduce contamination of plastic bags in the recycling stream, many US municipalities deem this effort insufficient, resulting in a flurry of bag bans and fees seeking to significantly change consumer behavior and decrease the use of plastic shopping bags.

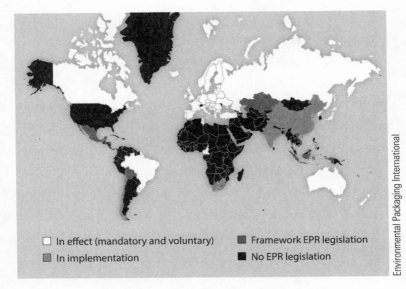

4.3 Countries with extended producer responsibility laws for packaging.

EPR laws that require brand owners to cover the cost of recycling post-consumer PPP provide an incentive to producers to reduce the amount of packaging they use, incorporate environmentally preferable materials into their packaging, and maximize material recovery and quality. In contrast to the fragmented municipal programs currently in place, well-designed EPR systems provide consistency by establishing statewide producer-funded programs that accept the same materials in all cities and towns and convey the same educational messaging.

Such policies also help meet the supply needs of industry. Today many brand owners that pledge to incorporate recycled content into their products often cannot procure enough recycled material to meet their needs. With strong

EPR laws, producers stand to gain access to greater amounts of post-consumer recycled material. These programs also offer financial incentives that encourage manufacturers to design their packaging to be more recyclable.

EPR packaging laws are spreading globally and growing in viability partly because the recycling or disposal cost is typically paid by manufacturers and their consumers, not taxpayers and government agencies, freeing up millions of dollars for other municipal services. In addition, these programs provide a direct financial incentive for manufacturers to use materials that are less expensive to recycle, increasing their value and opportunity to be brought back into the circular economy.

EPR packaging systems are continually evolving. The most innovative are those that charge a fee to manufacturers for each packaging material type based on its cost to recycle or dispose of. One such system charges manufacturers less for producing glass than plastics, as well as less for PET and HDPE containers, compared with films, polystyrene, and other plastics that are not easily recycled. This ***closed-loop recycling*** system provides a direct financial incentive for manufacturers to choose environmentally preferable (often more highly recyclable) materials in their packaging.

To be clear, all of this extra cost does directly end up in the price of the product a consumer pays in the end. But perhaps this cost is better incurred at checkout than in negative ***externalities***—like greenhouse gas emissions, marine debris, resource scarcity, toxicity, and food and drinking-water pollution—and continuing the burden on municipalities and taxpayers to subsidize waste.

The True Costs of Packaging

Technically, all packaging can be recycled so long as someone is willing to pay the cost; this is important to note when considering the commitments of various manufacturing companies to make their products technically recyclable by a certain date in the future. What makes a product or package *practically* recyclable is if it's economically viable to do so. As a hyperbolic example, if you want to make all of your products and packages recyclable, a simple answer is to make them from solid gold, as there would be intense competition to collect it. You may not even need garbage cans, as littered gold would be collected seconds after disposal.

In other words, making packaging out of materials with higher value (e.g., aluminum instead of plastic) will make them more likely to be collected and recycled. Inversely, if the cost to collect and recycle an item gets too high, perhaps because it is difficult to capture due to its small size (e.g., single-serve beverage pods) or its composition of multilayered material (e.g., flexible pouches), it becomes easier to dispose of it.

When it costs more to recycle than dispose, the extra cost to maintain recycling is shifted onto another stakeholder in the product's life. Taxpayers most often fund municipal programs in whole or in part, so the true cost of packaging ultimately falls on them (even if they never bought the product to begin with)—not the consumer who actually purchased it or the company that produced it. This is an inequitable allocation of costs since those who consume fewer products subsidize those who produce more waste. They also subsidize the industry that created the product and its packaging. Unlike in many other

Alexander Weickartt/Shutterstock

4.4 The true costs of a disposable coffee cup often fall on the consumer in the form of taxes and labor—not the manufacturer that produced the cup in the first place.

developed countries, in the United States manufacturers and brands are not responsible for their packaging once the consumer buys the product.

Take, for instance, the daily experience of a consumer purchasing a cup of coffee. If it is consumed inside a store, the retailer bears the cost of disposing of the associated garbage. If a person steps outside the store with the same paper coffee cup, gulps it down on the street, and dutifully places the cup in the public garbage can, the responsibility shifts to the municipality and, ultimately, the taxpayer. **SEE 4.4**

Do I Want EPR for Packaging?

Most of us don't like to be told what to do, especially when it comes to how we do business or how we spend our money, even if it's projected to benefit the environment or economy in the long run. Regulations and involuntary EPR requirements as a condition of sale can seem limiting and bad for business progress and profits; but when we look at the costs of the way we package, the benefits of ***product stewardship***—either compelled by government or taken up voluntarily—become quite interesting.

EPR laws are a mandatory type of product stewardship that includes, at a minimum, the requirement that the manufacturer's responsibility for its product extends to management of that product and its packaging when it hits the end of its life. There are two key features of this type of policy:

- Shifting financial and management responsibility, with government oversight, upstream to the manufacturer and away from the public sector

- Providing incentives to manufacturers to incorporate environmental considerations into the design of their products and packaging

The misconception about compulsory regulations of this type is that they force companies to behave in a certain way, an idea that irks big and small businesses alike. But with EPR no one is forcing corporations to use environmentally preferable materials; instead they are incentivizing them to do so. Some countries with EPR laws charge companies less for their production of environmentally preferable materials.

For example, in the Der Grüne Punkt (Green Dot) system used throughout Europe, the appearance of the trademark on packaging indicates that for such packaging a financial contribution has been paid to a national packaging recovery company for the cost to collect, sort, and recover the material. This system incentivizes companies to cut down on and choose economically viable materials to get the lowest fees. If a company chooses to stick to its unrecyclable design, that's fine—it just needs to pay accordingly.

In Germany packaging that is recyclable or made of *recyclates* or renewables gets a financial advantage by the PRO Europe (Packaging Recovery Organisation Europe) in a 2019 update to its packaging act. Experts say that this sort of packaging law is likely to be adopted by the European Union someday.

Implementing Extended Producer Responsibility

Improving the reuse, recycling, and remanufacturing of packaging is a big job that requires a lot of research, support, and expertise. Local, state, and federal government agencies have a lot to take care of with regard to putting plans for EPR laws into action:

- Research and analyze producer responsibility options, solutions, and implementation to inform and shape product stewardship policy

- Facilitate multistakeholder dialogues to drive consensus-based action plans

- Design and implement pilot *take-back programs*

- Draft bills, and track and analyze legislation

- Prepare and deliver testimony, and plan communications and outreach support

■ Plan enforcement strategies and department budgets

■ Ensure that producer programs are transparent and accountable to the public

■ Ensure a level playing field for all parties in the product value chain to maintain a competitive marketplace with open access to all

■ Set and enforce performance goals and standards—for supporting industry programs through procurement and for helping educate the public

Private-sector players such as industry associations, manufacturers, recyclers, and retailers have more flexibility. In general, the principles of product stewardship stipulate the following:

■ Programs should cover *all products in a given category,* including those from companies no longer in business and from companies that cannot be identified.

■ All producers within a particular product category *have the same requirements,* whether they choose to meet them individually or jointly with other producers.

■ Producers have flexibility to design the product management system to meet the performance goals established by government, with minimum government involvement. Producer-managed systems must *follow the resource conservation hierarchy of reduce, reuse, recycle, and beneficially use,* as appropriate.

■ Producer programs, including their development and the fate of the products managed, provide opportunity for input from all stakeholders.

So, what can businesses do now to lay the groundwork for EPR and reap the rewards of product stewardship? Small start-ups and large corporations can sponsor in-store and mail-in take-back programs for the consumer packaged goods they sell. Industry coalitions and other NGOs can start small with collection drives to send large shipments of material to organizations like TerraCycle for recycling, and then work up to something like the Carton Council of North America, which is a group of carton-based manufacturers that banded together to expand access to gable-top and aseptic carton recycling.

Who Should Institute Packaging Guidelines?

Although we are not yet at this level of *true cost accounting*, manufacturers in countries with systems that tie recycling cost to the fees they pay to place their packages on the market tend to make better decisions. Instead of selecting materials based only on upfront costs, they also base their selections on downstream costs of recycling or disposal.

All stakeholders—producers, governments, retailers, and consumers—have a role to play in an EPR packaging system, but the heart of EPR laws is the requirement that the manufacturer of the product pay for and manage its collection and recycling.

If a company sells 1 million pounds of packaging into a country, it should pay for that amount to be collected and

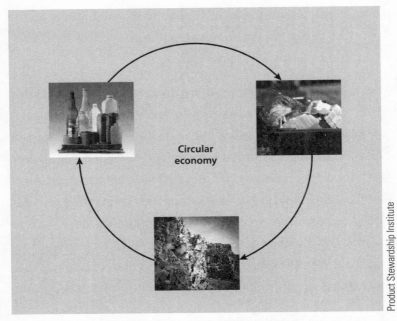

Product Stewardship Institute

4.5 A circular economy for packaging is possible if producers put up the financial resources to reclaim materials and engage governments and consumers.

recycled. Those fees can be set by a nonprofit ***stewardship organization,*** approved by a government oversight agency, that works on behalf of producers to maximize operational efficiency. Typically, such organizations contract for collection and recycling services, conduct education and outreach, report to the government oversight body, and determine the fees that each company must pay to the organization.

The government's role is to provide a level playing field for all producers, set system performance goals, and approve the stewardship organization's plan for collection, recycling, education, and the fees paid into the system.

Globally, manufacturers are increasingly being held primarily responsible for reducing the impact of their post-consumer packaging waste because they control the materials used. But providing take-back programs, investing in municipal recycling, and offering return incentives means little unless there is participation by consumers. **SEE 4.5**

While financial responsibility in EPR systems is best held by the producers, changing behavior requires all stakeholders—from retailers to consumers to government—to play a role. If those responsible are able to execute these initiatives and take the necessary steps toward EPR, consumers must use the collection systems provided for them by the producers, municipalities, and retailers. It takes a collective effort, ongoing communication, political will, and cultural acceptance. But the reward is meaningful and well worth the pursuit.

Recycled versus Recyclable

Stephen Sikra

Associate Director, Corporate R&D, Procter & Gamble

A DISCUSSION OF WHO *SHOULD* BE FINANCIALLY RESPONsible to ensure that packaging gets recycled must include an understanding of what makes packaging recyclable and why. With fragmented global waste management systems and an ever-evolving diversity of packaging technologies, recyclability can be confusing, even for the most well-intentioned person.

For a packaging designer, the most effective approach to take when considering materials is to make packaging out of material that recyclers want and have the technology to handle. That said, if someone is willing to pay for it, almost everything is *technically* recyclable. In other words, it's not solely the material composition or the condition of that material that determines recyclability; it's about the entire supply chain and the potential for a recycling company to make a profit.

For example, the lead author of this book, Tom Szaky, and his research-and-development team at TerraCycle have demonstrated through technology that some of world's most common, but unexpected, waste streams can be recycled; these include cigarette butts (among the most littered items on the planet), chewing gum (close behind), instrument strings, plastic food packaging, and even dirty diapers and used feminine hygiene products.

What Makes Something Recyclable?

The more processes required to integrate a waste material into a new product, the more it costs. To take the extreme example of dirty diapers, the technology required to properly clean, separate, and recycle them into usable materials is extensive and, as a result, expensive. In a hypothetical municipal recycling implementation, the combination of labor and processing technologies required to handle a dirty diaper at a facility as opposed to, say, an aluminum can, demands incremental and significant logistics and processing investment. This includes more training for personnel and a different set of sorting and processing technologies.

The delivery infrastructure for pickup and transport of soiled diapers would require special receptacles, storage units, and extensive education to make it easy, safe, and pleasant enough for the consumer to do properly. As a result, to bring diaper recycling to municipalities, there would need to be increased logistics and processing investments (both onetime and ongoing) and typically funds (perhaps from *voluntary producer responsibility* [*VPR*] programs or EPR-type legislation) to catalyze these initial investments and help satisfy the system "value equation." Whether or not a piece of packaging gets recycled is mainly a matter of economics. Recycling is a function of supply and demand for the secondary material in a market where it is cost-effective to do so. Someone has to be willing to pay for it, and if it costs too much money, it simply doesn't happen.

Supply and Demand: Recycled versus Recyclable

There are major material classes today that can be easily reintegrated into the same production cycle from whence they came. Plain, uncoated paper can be recycled into 100 percent recycled paper, clear glass can be almost endlessly recycled into new glass products, and aluminum cans be recycled into new ones over and over. Clear and rigid polyethylene terephthalate (#1 plastics) bottles can be recycled into new **PET** bottles, and high-density rigid polyethylene (#2 plastics) containers can be recycled into new **HDPE** containers.

Even with a recyclable design, many aspects affect the actual recycling of a product or package, with the primary factors being access and participation, separation, and end-market demand. According to the Association of Plastic Recyclers' "APR Design Guide for Plastics Recyclability," an item is recyclable in the United States if it meets the following conditions:

- At least 60 percent of consumers or communities have *access* to a collection system that accepts the item.

- It is most likely *sorted* correctly into a market-ready bale of a particular plastic meeting industry standard specifications through commonly used materials recovery systems, including single-stream and dual-stream **MRFs**, plastics recycling facilities, and systems that handle deposit containers, grocery store rigid plastic, and plastic films.

■ It can be cost-effectively further processed through a typical recycling system into a post-consumer plastic feedstock suitable for use in identifiable new products.[1]

Take clear PET: There is robust end-market demand for this material on the back end; because clear PET can be easily recycled into clear or colored iterations of PET, MRFs actively separate for it. But the high demand for clear PET is not enough to get it recycled. There are several steps between a discarded PET water bottle and its reintegration into the production cycle, starting with access to a collection program.

Access is determined in several ways, including whether a solution is convenient. A lack of appropriate receptacles in high-traffic spaces such as music venues, sports arenas, and many public thoroughfares, for example, will cause most consumers to drop a clear PET bottle in the nearest trash bin rather than hold on to it until they get home to their recycling bin. But no matter how many people have access to a recycling solution, they will not *participate* if they are not aware. Manufacturers have a responsibility in this supply/demand equation to improve awareness and access to recycling solutions for their products and packaging through education and on-pack messaging. **SEE 5.1**

Separation, beyond how it is collected (e.g., separate bins), describes the technical aspects within a MRF and other reclamation facilities. If a recyclable material gets to this point, contamination may still cause it to have a negative effect on the quality of the end product, and incorrect sorting could prevent it from being recycled at all. Businesses can invest in supply-chain security by working to support sortation and

5.1 Accepted-waste flyer from a municipal recycling program. Such resources engage consumers around recycling, improving participation by raising awareness.

processing at MRFs, where materials flow to both supply new production and divert from becoming waste.

But ***end-market demand*** is by far the most important factor in this equation. A strong pull for the material is needed to make the system work. Without a strong end market, "recyclability" can be diminished to nonexistence. Companies must consistently buy recycled material to maintain this demand.

		Access to collection	×	Supply Participation
Clear PET bottles	Japan, food grade	90%		90%
	European Union	85%		80%
	North America	67%		67%
Colored PET bottles	Japan, food grade	0%		0%
	European Union bottles	85%		80%
	North America bottles	67%		67%
HDPE bottles	European Union	85%		67%
	North America, natural	67%		67%
	North America, colored	67%		67%
Films	US store take-back	60%		1%
	US curbside (bag-in-bag)	1%		33%

5.2 The Association of Plastic Recyclers helped frame the recycling infrastructure into four components—access and participation, separation, and end-market demand—to come up with a conceptual recycling rate for different types of plastic.

The Association of Plastic Recyclers helped frame the recycling infrastructure into four components—access and participation, separation, and end-market demand—and turn

×	Separation	×	Demand	=	Recycling rate
			End markets		
	90%		100%		73%
	75%		100%		51%
	72%		100%		32%
	0%		0%		0%
	70%		30%		14%
	70%		50%		16%
	85%		40%		19%
	85%		100%		38%
	85%		90%		34%
	95%		90%		1%
	5%		90%		0.01%

it into an equation. This was extrapolated into a conceptual recycling rate for different types of plastic: clear PET bottles, colored PET bottles, HDPE bottles, and plastic films. Looking at a material and package form in this way enables us to understand how the recycling rate might be affected in any given situation. **SEE 5.2**

You can see that recycling colored PET gets tricky. This is because colored plastic cannot easily be turned back into clear plastic (like when a child mixes all their paints together, they can never be mechanically unmixed to their original hues—and the resulting mixed-color PET has lower end-market demand). As an industry, we need to find more end markets for colored PET. Chemical recycling (*depolymerization*) is one option to decolor *recycled PET*, or ***rPET***, but that certainly requires more infrastructure investment (more on that later).

Plastic films are even more difficult, as there is very little infrastructure for proper handling and separation at MRFs, and end-market demand is quite low for the resulting mixed plastics. Their collection is largely relegated to retail ***take-back programs***. As described in the previous chapter, what these programs have done is marginally reduce contamination at materials recovery facilities. Of course, consumers still throw their plastic shopping bags, bread and food sleeves, and bulk shrink wraps in both the trash and recycling bins—more on why such ***aspirational recycling***, or *wish-cycling*, is a problem in the next chapter.

Modern applications of plastic resins need to accommodate needs for flexibility and malleability; think containers for condiments, personal and beauty care goods, household products, and other items. Without ***plasticizers***—the small molecules added to polymers that push the polymer molecules slightly farther apart, making the material softer and more flexible—plastic would be too hard to interact with, as it would become fragile and could shatter with everyday use. Recycled or not, these additives to plastic make them difficult to recycle again.

Recycling Is the Preferred Solution

Recycling offers many environmental, social, and business benefits and supports a more circular production economy. The technology and processes exist today to get around almost any obstacle; as a result, we must strive to recycle even the challenging items, divert our packages from linear disposal through design, and increase end markets for recycled material.

The upside-down triangle illustration you see in figure 5.3 is a visual for the Waste Management Hierarchy developed by the European Union's Waste Framework Directive. **SEE 5.3** A similar hierarchy is used by the US Environmental Protection

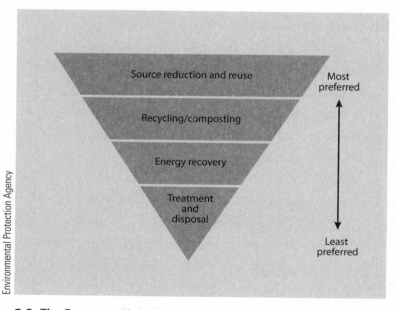

5.3 The European Union's Waste Management Hierarchy, also used by the US Environmental Protection Agency, ranks the strategies from most to least preferred in terms of environmental impact and efficiency.

Agency, ranking the strategies from most to least preferred in terms of environmental impact and efficiency,[2] placing *source reduction and reuse* up at the top.

Source reduction, or using less material, is a common practice for most businesses. Using less costs less, and a lighter, smaller package saves on material and transportation spends.

Reuse can translate into the creation or support of products that are durable and built to last, allowing consumers to use them repeatedly and for longer periods of time, offsetting the need for new materials and waste disposal with each cycle. This is the offset to **planned obsolescence,** the policy of planning or designing a product with a limited useful life so that it will become obsolete, that is, unfashionable or no longer functional after a certain period of time.

Consider a plastic fabric-softener bottle. Refilling the smaller, more functional bottle with fabric softener from a larger refill jug enables reuse of the small bottle and its features. Reuse eliminates the need for new energy and resources to manufacture another bottle. This route concludes when an object is ultimately discarded.

Recycling converts discarded items into new, usable materials. Many items can be recycled multiple times, with glass and metals being the most recyclable, then plastics, and finally fibers. **Composting,** also known as *organics recycling,* describes how organic materials (such as food scraps) are converted into energy or **compost** (for at least a second life). Recycling is lower on the hierarchy due to the energy and resources required to convert discarded items into new, usable **inputs** for production.

Recycling is preferred to *energy recovery*—in which waste material is incinerated for the purpose of energy capture—which is preferred to landfilling, or burying waste between layers of earth. While we strive to be at the top of the model and prefer recycling to *landfilling*, a well-managed landfill is better than *litter*, particularly litter in our marine systems.

As designers, choose to work up the hierarchy by using materials that are reusable or recyclable today while advancing the recycling infrastructure for more-complex materials and packaging configurations.

Designing to Improve the Market for Recycled Materials

Recycled materials are those that are collected, processed, and purchased for use back in new production. *Recycled content* refers to the percentage of this recycled material in a new product or package. Choosing to integrate recycled content in your packaging will support the market for recycled materials.

When designing a product, consideration must be given to function, appearance, and cost. Packages must perform their intended tasks of delivery, protection, branding, and additional functionality (such as pourability and reseal technologies), and package designers are tasked to ensure that recycled materials will deliver as needed and within tight cost considerations. With advances in physical and mechanical sorting and optical and *NIR* (*near-infrared*) detection systems, obtaining high-quality PCR material is an option.

Chemical tracer and digital watermark technologies are currently in development to further improve sortation for the creation of a high-quality end product. There is a lot of development potential to creating a "barcode of recycling," with the possibility to reduce the amount of sorting required by consumers and aid the sorting of plastics at MRFs and recycling plants.

Using PCR materials can present some challenges with color, aesthetics, and feel, which are important and almost always part of a brand's identity. Chemical recycling technologies have proven the ability to remove pigments, dyes, and additives to produce "virgin-like resins" that are competitive with virgin raw material. Procter & Gamble (P&G) recently invented the *PureCycle Process* and licensed it to PureCycle Technologies to open a plant to restore used *polypropylene* (*PP*, or #5 plastics) plastic to "virgin-like" quality[3] and remove colors and contaminants from *rPP*.

It is very important to note that the inclusion of PCR materials in the production of a new item *may* have an impact on the subsequent recyclability of that item. End-market demand is based on numerous factors, but most important are *composition* and *contamination*. In situations where a high-quality PCR material of the same composition as the parent package (a PET bottle with rPET PCR) is used, the PCR material has no impact on recyclability. Whether the bottle contains 10 percent PCR or 100 percent PCR, recyclability is (generally) not affected.

If the chosen material introduces a mixed resin (generic recycled content often has little traceability) into a typical monomaterial application (such as adding recycled material from flexible film containing polyethylene and PET layers into a polypropylene rigid part), however, the new part is considered a mix of resins. This mixed recycled resin will have less end-market demand. As a result, its recyclability is certainly limited and, in most cases, restricted to mixed-resin applications. In today's market for recycled content, a monomaterial resin has a larger end market than a multimaterial resin due to the current capabilities of MRFs.

Storied plastics are resins collected by waste stream, sorted by material type, and traced back to the point of origin (e.g., "recycled ocean plastic" or "recycled cosmetics packaging"); they can help brands circumvent some of the prohibitive barriers of cost, quality, and diminished recyclability for mixed recycled resins. Companies like TerraCycle specialize in the capture and collection of common but difficult-to-recycle streams—like ocean plastic, personal and beauty care packaging, juice and snack pouches, water filters, and more—putting them in the position to supply to manufacturers dozens of different storied plastics sourced from their collection streams. **SEE 5.4**

Storied plastics can offer traceability and offset the diminished physical material quality of standard PCR material (rPET or rHDPE). Incentivized by increased access to a secure supply chain of PCR material, which creates interest for consumers, companies can work to increase consumer access to recycling collections and inspire participation in the programs through marketing, industry associations, and on-pack messaging (the How2Recycle label is a good example). Many **CPG** companies today are using PCR materials, but using storied plastics provides the exceptional opportunity for manufacturers and major

5.4 Storied plastics are resins collected by waste stream, sorted by material type, and traced back to the original point of origin, such as "recycled cosmetics packaging."

brands to differentiate, carrying a narrative that consumers can connect with and would be willing to pay a premium for.

Designing a highly recyclable package out of recycled material is essential to advancing the recycling infrastructure and reducing waste from both ends. Ideally, a package is one that can be recycled back into itself; this concept is called ***closed-loop recycling***. In contrast to ***open-loop recycling***, which is where a material is recycled into other types of products (e.g., soda bottle into fiber), items that fall into the closed-loop recycling scheme pass what we call the ***Circular Design Test***. A package or product is a winner if it can be circulated from production to consumption and back again, over and over.

At P&G we are blessed with many like-minded partners participating in collaborative, pre-competitive efforts. In 2018 P&G celebrated its thirtieth year of using PCR in its packaging, thanks to global partners Plastipak, Cleantech Group, Envision, KW Plastics, and many others. Consistent, year-on-year use of secondary material in increasing amounts helps maintain and advance the recycling infrastructure. We are happy to do our part—but we know that no one company can do this alone.

How Do I Ensure That My Package Is Recyclable?

Diligence in design is necessary. Weight reduction, aligning with the very top of the waste hierarchy, is a great start. But advocating for recyclability and the inclusion of PCR materials in your package can help you find new solutions and add up to the best sustainability profile. Here are some tips and tricks for how to design the most sustainable package possible:

- Call your local recycler (or MRF) and ask what they want to see in their waste stream. Don't assume to know what is recyclable, or marketable, in each recycling system. Ask the question: *Would you want this in your stream*? Not: *Can you technically recycle it?*

- Use highly recyclable clear glass, aluminum, simple paper, and rigid and clear PET or HDPE plastics in your package design.

■ Use single-polymer, monomaterial configurations that don't require separation. Try to avoid multipolymer materials.

■ Use clear and rigid plastic when possible. If colored or opaque, find ways to reuse colored recycled plastic in your package.

■ Implement on-pack recycling messaging with instructions, such as the How2Recycle label.

■ If using a label for on-pack design and messaging, consider a perforated label rather than printing colored plastic. For example, P&G's Lenor fabric enhancer in highly recyclable PET bottles is wrapped in a perforated sleeve label with marketing and content messaging, with instructions on separating the components for recycling.

■ Know which coatings and additives (such as moisture barriers or colorants) reduce end-market demand. Work to improve technologies to handle these items.

■ Know where your recycled content comes from and communicate its origin to consumers. Not only does this offer traceability, but consumers will be more likely to recycle it.

■ Support recycling systems and emerging technologies. For example, P&G is collaborating with Materials Recovery for the Future[4] to demonstrate flexible-film recycling in a MRF at scale and with The Recycling

Partnership[5] and many other collaboratives to advance recycling in local communities.

Recycling is a function of supply and demand. Designing for a circular economy means not only creating a package that fits into the current system but working to improve that system. Work to increase end markets by purchasing recycled material. Strive to create the most useful design with life-cycle thinking, and support source reduction and the reuse of valuable materials. Look to collaborate in the development of a comprehensive recycling infrastructure and, in the meantime, play your part.

CHAPTER 6

Designing Packaging for the Simple Recycler: How MRFs Work

Ron Gonen

Cofounder and Chief Executive Officer, Closed Loop
Partners; Cofounder and Former CEO, Recyclebank;
Former Deputy Commissioner for Recycling and
Sustainability, New York City Department of Sanitation

THE ECONOMIC INCENTIVE TO COLLECT, PROCESS, AND sell post-consumer waste for use in new production comes from whether or not recyclers can make a profit on the sale of materials after their costs to collect and process it. These costs are in part related to the capabilities of the local *materials recovery facility*, or **MRF**.[1] Also known as a *materials reclamation facility, materials recycling facility,* and *multi-reuse facility,* a MRF is the intermediate processing destination for potentially recyclable goods once they are collected from a home or business.

MRFs are specialized facilities that receive, separate, and prepare waste materials for sale to end-user manufacturing companies. They operate using a combination of equipment and manual labor. A MRF can be thought of as a large machine through which waste flows before it comes out the other end as usable raw material. The thing about machines is that they are mechanically designed to perform a finite array of tasks.

In the case of the MRF, this means it can accept and process only materials that are considered recyclable. Items that are profitable for waste management companies to recycle in the specific regional market are the ones on which the MRF is built to focus.

Products and packaging that are considered unrecyclable are incompatible with the technology and manpower implemented to process the profitable waste streams; they are not profitable for waste management companies to recycle and thus contaminate the recyclables stream. These items tend to be problematic for MRFs, and they can jam up, slow down, and sustain damage (to equipment and workers). Ultimately wasting time, money, and valuable resources, these items not only do not get recycled but they diminish the quality and financial value of the materials that are being focused on for recycling.

For example, clear *PET* (the plastic your soda bottle is made from) is very valuable, but when colored, semitranslucent, or opaque PET gets mixed in, it contaminates the clear PET, reducing its value.

Man and Machine: The First MRF

Understanding how MRFs work can provide some insight into why certain items are accepted for recycling and others are not. The first modern recycling facility of this kind in the United States appeared in the 1970s, when Peter Karter's Resource Recovery Systems went into business. After some time as an engineer specializing in the disposal of nuclear waste in the years following World War II, Karter started enlisting volunteers to collect glass bottles, which he and his team, clad in goggles and gloves, separated by color and provided to a glass company for the production of new glass.

By this time many communities had set up unofficial recycling programs in which residents drove their cans and bottles to a recycling depot. In the wake of the first Earth Day on April 22, 1970, there was growing interest in recycling but no efficient way to process glass and metal. Recycling was still an expensive service that towns were reluctant to pay for, and it was difficult to find a market for the materials.

Karter's wife, Elizabeth, challenged her husband to come up with a more efficient system. He then invented and patented several machines that sorted glass by color and cans by metal content. A neighbor, Mervin Roberts, invented a machine that used light beams (the precursor to today's infrared, optical scanners or near-infrared [NIR] machines) to sort glass. Eventually, the company developed methods to process plastic and cardboard and, ultimately, replicable systems for mixed recyclables. The technical problems overcome by these inventions made recycling more economically viable with systems that were sophisticated enough to sort and separate high volumes of material.

How MRFs Work

Today's MRFs generally work as intermediaries between their operators and the respective municipality. The municipality is motivated to recycle as much as possible, as the benefit is the avoidance of landfill disposal fees. Meanwhile the MRF is motivated to accept only those items that can both be recycled and sold at a profit. This negotiation is what determines what is collected for recycling.

As discussed in the previous chapter, someone always has to pay to recycle or dispose of municipal solid waste, and it's almost always the municipality and its taxpayers. A MRF or other centralized processing facility is planned and designed to recycle to the extent that it plays a role in a jurisdiction's integrated waste management plan. Thus every MRF is different because every community's recycling program is different. **SEE 6.1**

A tour of a single-stream recycling plant illustrates the obstacle course that is the sorting process. After collection, residential and commercial waste is brought to the MRF by truck. It is first dumped out on a tipping floor, a large recessed area, to be pushed into a large feeding container using a front-end loader. Experts call this approach one of the most inefficient components of the MRF because dropping and moving material on the floor requires additional equipment and causes large amounts of glass breakage.[2]

Then the feeder inclines to the first conveyor for a presort, where workers manually remove large bulky items and other contaminants. Consumers think many items ought to be accepted for recycling when they are not, so this presort has been known to extract such objects as car parts, bicycles, 5-gallon pails, garden hoses, working smartphones and laptops, and even an actual German Enigma machine from World War II.[3]

Ideally, it is also here that unrecyclable items such as plastic bags, flexible pouches, multi-compositional containers,

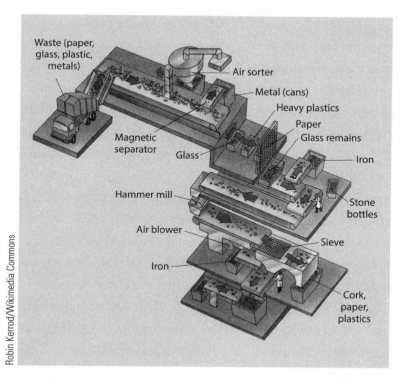

Robin Kerrod/Wikimedia Commons

6.1 This diagram of a single-stream recycling plant illustrates the obstacle course a recyclable must get through to have a chance at being used in new production.

and natural and synthetic combinations (e.g., paper coffee cup and plastic top) are picked out manually. But contaminants, large and small, still make their way through.

From presort, the material moves on to a series of moving and vibrating screens that separate the waste by material type. Some MRFs shred the waste at this stage to ensure an even feed and spread of materials across the belts in the whole plant and to aid the advanced optical technologies employed to view and sort the waste efficiently.[4]

Other MRFs use an **OCC screen,** where smaller, pliable material falls though turning star-shaped gears. It imparts a bouncing, wavelike action on the material stream, and larger rigid objects, like cardboard, bounce over the top. The smaller material drops onto a fine screen to break and remove glass and other 2-inch or smaller material. The broken glass moves to a glass cleanup system for further separation to make it clean and sellable. **SEE 6.2**

These small-form items are one of three parts through which the technologies tend to sort material. The other two parts focus on the two-dimensional items (paper and card-board—about 50 percent of the stream) and three-dimensional items (plastic bottles and jugs, and aluminum and metal cans). Once the fiber (paper) is recovered, 3D objects bounce back and are conveyed to an optical sorter for final fiber removal; optical scanners are also used to sort the different plastic resin types into bales.

Remaining containers travel under a magnet to lift ferrous metals and cans from the transfer conveyor into a storage bin for baling. Any lingering materials travel on a type of separator, which propels aluminum cans over a separation gate to the appropriate bin.

Innovations of technology provide some solution to the fallibility of human labor in picking and sorting out contaminants, as well as the hazards this work poses. Pilot robotics programs are using combinations of computer vision, machine learning, and artificial intelligence to run synchronized robotic arms to identify, sort, and pick recycled materials from moving conveyor belts.[5] It's a glimpse into the future of smarter,

Bulk Handling Systems

6.2 The turning gears of an OCC screen are among the many tools in a series of methods that MRFs use to sort materials and prepare them for recycling.

more streamlined recycling. Of course, these technologies are expensive, so scaling up on these to expand the recyclables streams will take time. **SEE 6.3**

Contamination: Waste of Time, Money, and Resources

The journey of a recyclable—from a bin to a bale—is complex. It is one that a piece of waste will complete only when

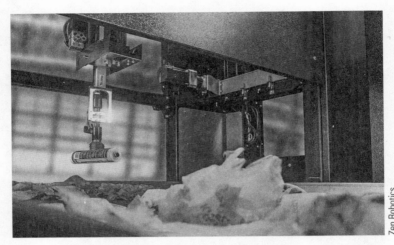

Zen Robotics

6.3 Are robots the future of expanding recyclables streams? Perhaps, but building up and investing in these programs will take time and money.

it is designed to flow smoothly into and through the system's capabilities. Even then, improperly sorted waste items can contaminate it.

The best reasonable case scenario in the MRF is when all nonorganic ***contaminants***—like plastic bags, Styrofoam, and flexible packaging—are eliminated from the waste stream. The next best is when these are removed via manual or mechanical sortation. The absolute best-case scenario? Eliminating these materials from the economy entirely. **SEE 6.4**

Besides potentially damaging machinery and losing the commercial and practical value of material to linear disposal, contaminants also divert manpower away from quality control. Rogue unrecyclables that shouldn't have been placed in the recycling bin in the first place do find their way into the finished materials.

6.4 Contaminants that do make it to a MRF are, one hopes, picked out in presort, but it would be better if these items were eliminated from the economy.

Typically, MRFs can sell the following post-consumer recyclables:

- Uncoated cardboard

- Uncoated paper, like photocopy paper and newspaper

- Rigid and clear PET plastic (beverage bottles), or #1 plastics

- Rigid and white HDPE plastic (jugs), or #2 plastics

- Rigid PP (shampoo bottles), or #5 plastics

- Aluminum

- Clear glass

Almost all materials not on this list fall outside the scope of recyclability of most MRFs. The exceptions are few, and there is really no way of knowing them for sure, save calling the MRF and asking.

Call Your Local MRF

As discussed in the previous chapter, recyclability is determined by supply and demand. The most effective way to make a recyclable package (supply) is to design it out of material that recyclers want (demand). If recyclers want it, they will have the processes to attain it in the form of collection systems and facility capacities. Because the capabilities of municipal recycling facilities vary from region to region, to be considered "highly recyclable" your package needs to fit into all or most of them.

Global *CPG* corporations likely have distribution across the range of international markets. Committing a team to gather recycling capability data on even the top, largest markets provides useful information about the "general" recyclability of your package. Small businesses and start-ups setting out to choose from a catalog of packaging options to purchase should do the same.

Don't make assumptions: call the recycling facilities in the markets where your package will be consumed to get the most informed, clear idea of the value of your materials. Talking to the plants local to your production hub, office, or manufacturing facility isn't enough to know if your package is accepted for recycling where it will be used. Start with the largest recyclers in your product's distribution markets. Out of those, call the simplest recyclers, or the ones that accept the smallest number of waste streams, to find out what they actually want (not what they can technically recycle).

The questions you should ask MRFs don't stop at *Would you, and could you, accept this item for recycling?* Information on the MRF's website and in its communications will tell you that. The key to assessing the recyclability of your package comes with the answers to these questions: *How much would you pay for it?* and *Would you want this waste stream in your recycling bin?* This information determines how valuable your package is in the market for post-consumer waste.

Recyclers are in the business of providing high-quality, marketable PCR materials to manufacturers and companies; they are a business like any other and prefer more-profitable materials over ones that are less so. They view all materials through this lens, and they value those that are most profitable to them. More than that, contaminants and recyclable materials with a low-end market become a cost to dispose of.

Considering the recyclability of your package from the perspective of end markets is the most effective way to determine how it fits within the recycling system.

Design Simply and Avoid "Watch-Outs"

When designing products, design for the capabilities of the average MRF versus the most advanced. All MRFs are different, comprising a large matrix of what can be recycled municipally. Looking at the simplest MRFs with the most basic capabilities can guide the design of a package or product and allow it to enjoy the highest rate of recycling across the marketplace.

The following simple examples illustrate recycling "watch-outs" of which many designers are unaware.

■ **Avoid black and dark plastics** (even if the resin
   number on the bottom is accepted by the local municipality). The optical scanners used to identify types of
   plastic at MRFs using the reflection of light deem black
   plastic unrecyclable in the current infrastructure. This
   is because black plastic does not reflect light. Thus the
   rigid black plastic of microwave food trays, takeout
   containers, and other items is not accepted by most
   MRFs. **SEE 6.5**

■ **Avoid flexible-plastic films,** as they are of particular
   difficulty at most MRFs. Falling into this category,
   along with plastic shopping bags, are the thin plastic
   films used to encase bread products, paper towels,
   napkins, bathroom tissue, diapers, cases of bottled
   water, fresh produce, and much more. While rigid and
   flexible plastic may shred and fall through, plastic films
   can jam up the cogs and conveyors at MRFs, wrapping
   around gears and getting stuck.

■ **Avoid multimaterial packaging configurations,** a
   recycling "don't" due to the need for separation at the
   material level. Such multilayered flexible films include
   paper blister packs with foil, single-serve beverage
   pods (made of plastic, metal, and organics), and paper
   cartons with plastic fitments. If these are not manually separated out by consumers, the MRF sends them
   straight to landfill; workers will not take the time to
   separate these items. Further, because of the small size
   of the various units, like caps and plastic fitments, they

6.5 These are resin numbers—not "recycling" numbers. Not all packages with these numbers can be recycled, and some have variables that make them unrecyclable.

would fall through the cracks and outside the scope of capture.

Keep packaging and product design simple. One point to keep in mind is that some of a package's most visually interesting or appealing aspects (like multimaterial inclusions) are what render it hard to recycle. As mentioned earlier, colored plastic cannot be made into a lighter color or clear, so colored messaging and designs on plastic render it either difficult or impossible to recycle. Opaque PET has no market for the same reason and tends not to be recycled. Glass is one of the most highly recyclable materials, but the same is true for colored and gradient glass (glass that starts clear and transitions to color).

A way to understand the issue of color and end markets with both glass and plastic is to consider household paint. Paint always starts out white; when a color is chosen, the retailer adds in the dyes and pigments to get the desired shade. After the paint has been mixed, it cannot get back to white. The only place to go from there is to add more color, yielding a darker outcome—ultimately turning black. This is why retailers sell paint that has been mixed at a deep discount to white paint.

In the world of glass and plastic, clear is like white paint; it has great value for its potential in many applications as opposed to only a few. Colored glass and opaque or colored plastic is like paint that has been mixed into a color you don't want; in many cases, it simply becomes waste.

Action for Retailers, Governments, and Consumers

Basically, the best way to enable a recyclable outcome within the currently fragmented infrastructure is to design and buy with clear or white single-compositional packaging composed of highly marketable, sellable materials.

While packaging designers, purchasers, and businesses have an obligation to prefer packaging acceptable to MRFs, retailers should support them by choosing to stock products that bring this type of packaging to market versus ones that don't. Many stores already dedicate themselves to selling fair-trade, eco-friendly, clean products—and they command a premium for them. Retail can do the same with low-impact, recyclable packaging and use communications to show value and drive sales.

In the same way that recycling is a function of supply and demand, recyclable products and packaging must supply a demand in the marketplace. Governments can provide incentives to manufacturers that design environmentally preferable packaging through **extended producer responsibility (EPR)** laws and by extending them to retailers that stock a percentage of preferably packaged items. Governments can also provide educational literature (such as a simple "accepted waste" poster), cost savings for waste diversion (aka "pay as you throw" systems), and other enforcement methods.

Cities and municipalities can support local and small businesses in their mission to integrate sustainability into both their packaging and their processes. Through incentives, such as money savings from sending less to landfill, businesses can be encouraged to operate with a focus on recyclability. Perhaps cities can require municipal solid waste training for new businesses to acclimate them to the local recycling system.

Consumers need to participate, educate themselves about what is actually accepted for recycling, and be vigilant about recycling correctly:

- Plastic bags full of recyclables go straight in the garbage.

- Recyclables don't have to be squeaky clean, but if you wouldn't want to handle it, it will probably get tossed.

- Despite the **resin identification code** on the package, black, dark, and multicolored plastic is not recyclable.

- The same goes for gradient glass.

- Just because a package says "Made with recycled content" doesn't mean the package is recyclable.

■ **Bioplastics**—plastics derived from renewable biomass sources—tend not to be recyclable despite the triangle (more on that in the next chapter).

■ Natural and synthetic packaging combos, such as a paper coffee cup with a plastic lid—or the plastic, metal, and organic combos of single-serve coffee pods—are not recyclable.

■ Calling your local recycler will give you the best information about what belongs in your bin and what doesn't.

More important than being aware of how your local MRF accepts waste is applying that knowledge through active participation in the recycling system. Even more important than that is changing your consumption to favor products and packaging made of highly recyclable material, as well as items that are more reusable, durable, and long-lasting. Supporting brands that are doing good work is essential to driving value for more-circular business practices.

Supporting the development of new infrastructures and improved systems at MRFs is critical, but the solution must come from packaging design: by designing simply and buying simply, harking back to the material makeup of packaging before more-complex materials arrived. From the packaging producer to the local waste management company doing the work, striving to flow within the current system will ensure that your package is kept at its highest value at all times.

The Myth of Biodegradability

Mike Manna

Founder and Managing Director,
Organic Recycling Solutions

MANY PACKAGING MATERIALS FAVORED FOR FUNCTION and cost don't fit into the capabilities of today's recycling facilities. As a result, the potential to convert them into secondary feedstocks are lost, and they end up in the landfill, the incinerator, or floating around as litter.

In a hunt to innovate out of this challenge, packaging designers and plastics producers are in the market for materials that replace oil-based ones and more readily degrade in the environment. These could ease the strain on fossil fuels and at the same time reduce plastics pollution—both of which directly and indirectly have a detrimental effect on the oceans, wildlife, and public health. Thus supplying this demand for sustainability and resource efficiency in packaging is a growing market for bioplastics.

Bioplastics and Biodegradable Plastics: The Next Frontier?

Bioplastics are plastics in which all carbon is derived from natural, renewable feedstocks. These components can include

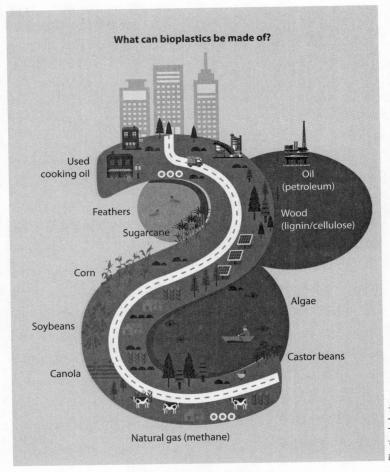

7.1 An alternative to petroleum-based resins, bioplastics are derived from natural, renewable feedstocks, which include a variety of sources, including corn and algae.

corn, potatoes, rice, tapioca, palm fiber, wood cellulose, wheat fiber, and sugar; they are renewable in the sense that they can be grown and regenerated again and again. Shrimp shells, seaweed, and algae have also been used as a base for bioplastics. **SEE 7.1**

Packaging is currently the largest application of bioplastics, accounting for 40 percent of the market.[1] The bioplastics industry has a current global production capacity of 3.8 million metric tons; it is predicted to grow an additional 50 percent and be worth around $7.2 billion by 2022.[2] Some experts believe that it could possibly replace up to 90 percent of traditional fossil fuel–based plastics in the future.

The thought is that plastics made with plants, as opposed to fossil fuels, will sustain the unstoppable trajectory of the world's consumption with a more sustainable material. The material is already delivering in the consumer durables sector, being used in automotive parts, furniture, eyeglass frames, and such cutting-edge applications as 3D printing and medical implants.[3]

For single-use disposable plastic packaging items in the consumer goods space, biodegradable bioplastics in particular promise long-term advantages.

How Biodegradable Are Bioplastics?

The advantages of *biodegradable* bioplastics hinge on two key factors:

- ■ The raw materials used to produce them are more sustainably sourced than petroleum-based plastic.

■ Consumers and manufacturers would no longer have
to worry about where these items end up, as they would
degrade naturally—like a banana peel.

The latter factor, however, has mobilized a torrent of mis-
information, misplaced optimism, consumer confusion, and
headaches for recyclers and composters alike.

For context, traditional plastics that are made from fossil
fuels are not biodegradable, but they are *degradable*. Everything
degrades *eventually*, even if it takes a thousand years, so the
term *biodegradable* can be vague and misleading.

Sometimes petroleum-based polymers, known as
oxo-biodegradable plastics, are combined with biodegradable additives to facilitate degradation triggered by ultraviolet
(UV) radiation or heat. This subcategory of plastics, however,
does not decompose into harmless residues that are absorbed
back into the environment. Instead they are designed to more
rapidly break apart, simply turning into harmful **microplastics**
that end up in our food and drinking water. **SEE 7.2**

Microplastics are ingested by all life, from the biggest
whales to the littlest larvae, contaminating and poisoning our
ecosystem. Some of the world's best-known brands have pub-
licly denounced the use of oxo-biodegradable plastic for these
reasons,[4] in addition to the fact that the bioadditives render
the material absolutely unrecyclable due to their instability.

In contrast, **biodegradable bioplastics** are intended to
behave exactly as they sound: when discarded, the material
will completely biologically break down into carbon dioxide,
water, and biomass—elements that can be readily digested by

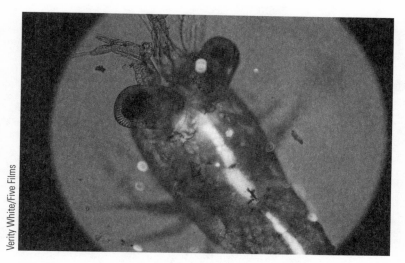

7.2 Plankton ingesting microplastics, here dyed to glow brightly. Microplastics work their way up the food chain and are eaten by organisms big and small, including humans.

the surrounding environment. ***Compostable bioplastics*** are the subset of these that will break down in a compost site "at a rate consistent with other known, compostable materials,"[5] such as paper, food waste, and yard trimmings.

Consumer Confusion

Popular applications of bioplastic bearing the "biodegradable" and "compostable" labels are single-use items such as plastic cutlery, to-go meal containers, beverage cups and straws, and plastic shopping and produce bags. Consumers mistakenly connect with the idea of a disposable plastic that will dissolve into nothing instead of floating around in the ocean forever, and they often think of such items as a way to easily reduce their impact.

What most consumers don't realize is that biodegradable bioplastics will break down only under the right conditions—those of an industrial composting facility. And even if that happens, they won't contribute value to the compost, unlike coffee grounds or leaves, which have a wide range of micro- and macronutrients as well as a living ecosystem of bacteria and other microbes. High temperatures, moisture, and exposure to UV light need to be strictly regulated for these materials to break down. Even in these facilities, some biodegradable plastics can take up to a year to be fully processed. **SEE 7.3**

This isn't a job for a natural environment like a forest or a beach—or even your small-batch, backyard compost bin: it's a job for a very specific set of processes that occur only if the right microbes are in the right place at the right time.

7.3 "Compostable" corn-based plastic cups after two years in an active home compost pile that reached 158 degrees F.

Zane Selvans/Flickr

Even an orange peel or a banana[6] will mummify in a landfill, and according to a report by the United Nations Environment Programme, biodegradable plastics that find their way into oceans and waterways have a big chance of not biodegrading.[7] The same goes for if they're tossed out a car window, dropped on a hiking trail, or left on the seashore. And we have to ask ourselves: *Does putting "biodegradable" on a package encourage littering?*

Despite the existence of composting solutions, 45 million metric tons of organic waste still end up in landfills across the United States each year. With about 200 industrial composting facilities (serving less than 5 percent of the US population), it's safe to say that we are currently ill-equipped to compost meaningful volumes of food and yard waste, let alone bio-degradable plastic.

In the same way that everything is technically recyclable, authentic biodegradable bioplastics may technically eventually break down into organic elements; but if the general population doesn't have access to a solution, these descriptors become meaningless.

A Conundrum for Composters

Of the very few composters in the country, even fewer accept bioplastic for recycling with organics. Composters are a business, and their goal is to make rich compost that is full of micro- and macronutrients and a full ecosystem of beneficial bacteria and microbes. To reduce contamination and strive for a high-quality product, recyclers will not accept items that do not fit into their capabilities; the same goes for composters.

In a sense, both traditional recyclers and organics recyclers serve the same function: to process discarded materials and turn them into a useful *input* for new production. Although biodegradable plastic will technically degrade under the right conditions, it doesn't yield good compost and is viewed as a contaminant as opposed to a useful input. This is why composters typically don't want it, even if it can degrade properly.

Here is a simple way to think about composting: if you were presented with a choice to eat either a beautiful sandwich filled with all sorts of vegetables and other yummy things, or a piece of cardboard, which would you eat? For a composter, a sandwich is nutrient-rich food waste like coffee grounds and yard trimmings, and cardboard is the bioplastic. The latter is technically "edible," but it is heavily processed and deprived of any organic goodness that will help plants grow and thrive. For all intents and purposes, both a sandwich and a piece of cardboard would digest and pass through your system and come out the other end; in the case of the composter, both materials would result in compost. But in both scenarios, the former is more desirable than the latter for a healthy result.

A genuine biodegradable plastic will be converted to carbon dioxide, water, and compost without leaving any persistent or toxic residue, maintaining the integrity of the resulting compost. The by-products of some biodegradable plastic compositions cannot be confirmed, however;[8] thus most composters will not take any biodegradable or compostable bioplastics to begin with, considering them contaminants to the end product of organics recycling.

Confirming Compostability

For a bioplastic to be considered recyclable, most consumers must have access to a recycling solution (in this case, for organics). Given the woefully inadequate number of facilities in the United States alone, one could argue that every biodegradable or compostable bioplastic is not organically recyclable.

That being said, the composters that do exist have standards for the materials they accept for integration in their compost mix. As with a traditional materials recovery facility, a composting company is built to handle only the materials that they consider desirable ingredients for an end product that producers want. Composters want their customers—often farms and garden organizations—to continue buying their compost.

Contacting a range of composters in different markets and with different client bases will give you the best feedback on the acceptability of your packaging material. Every organics recycler is different. Those that receive compostables from only households and small businesses, as opposed to large food producers or paper manufacturers, will give you a different answer. To whom composters sell their finished product will also influence their outlook.

Don't stop your line of questioning at "Will this degrade in your composting plant?" or "Can you accept this item for composting?" Not only will most composting websites tell you what you want to hear but most people on the phone will tell you the same. While a composting facility's first order is making great compost, its second is touting its capabilities. Remember, like plastics recyclers, composters want to project themselves as advanced and sophisticated, so they will likely almost always say that they can *technically* process something.

Instead, ask: "How much would you pay for it?" and "Would you want this in your supply chain?" This shows how valuable the type of bioplastic you use in your package is as an input to a marketable compost product. It may also prompt a more honest answer about the composter's capabilities. Remember that what a recycler considers a contaminant to its end product represents simply a disposal expense.

To complicate matters, just because a plastic is sourced from renewable biomass doesn't mean it is biodegradable. Bioplastics as a whole are often conflated with biodegradability in a similar way that recycled content in a package is often associated with recyclability (note that they have nothing to do with each other). All biodegradable plastics are biobased, but not all bioplastics are biodegradable.

Is There Such a Thing as a "Green Plastic"?

As is often the case in the world of sustainability, there is much more to this conversation than most consumers are aware of. Misleading labels and incongruent information with regard to the amount of sustainably sourced content in a package is one aspect.

To be called a "bioplastic," a material need only be composed of a percentage of renewable material. A majority percentage could very well be fossil fuel–based plastic resins and synthetic additives. In the United States, this is defined on a product-by-product basis via certification under the US Department of Agriculture (USDA) BioPreferred Program, a federal program that promotes the purchase and use of biobased products through a labeling initiative similar to that of the USDA National Organic Program.[9]

Like the USDA organic certification, an indicator of biobased plastic content offers businesses the opportunity to command a premium in the market for sustainable, high-quality goods. We as consumers connect with the concept of a plastic made from plants. From a marketing perspective, the "bioplastic" label taps into an increasingly conscientious consumer base. That a package labeled "bioplastic" may contain only a small fraction of sustainably sourced material is misleading and is an example of *greenwashing*—a form of spin in which "green" marketing is deceptively used to promote the perception that an organization's products, aims, or policies are environmentally friendly when they are not.

The viability of bioplastics as a more sustainably sourced, resource-efficient material is also questionable. Back in the early 2000s when oil prices were much higher and there was a demand for biofuels from companies planting corn, sunflowers, sugarcane, and switchgrass, there were problems with food shortages.[10] Diverting a large amount of agricultural capacity to produce bioplastics now would have consequences for food production, land use, conservation, and biodiversity protection.

Humanity has pretty much maxed out agricultural land. Offsetting demand for petroleum-derived plastics with plant-derived bioplastics would call for millions of additional acres of agricultural space. Already-threatened rain forest regions such as South America, where the tropical climate is necessary to grow feedstock such as sugarcane, would be especially at risk. Recent developments in the world of vertical farming could make sourcing renewable feedstocks less of an issue, but agriculture has a significant water footprint no matter how you slice it.

With biodegradable bioplastics, all the precious resources and energy required to produce them effectively go to waste after one use. Although they will degrade, they won't degrade into anything like the nutrient-rich earth at the start of the process. It just doesn't make environmental sense to take a plant, turn that plant into a highly refined petrochemical, only to then use it once and have it turn into something effectively worse than soil. There is potential for compostable plastics made from waste products, such as fruit pulp, wood shavings, and sewage, but to produce an economically viable supply, the infrastructure needs to be developed. **SEE 7.4**

On top of it all, most bioplastics, especially those labeled "biodegradable," are not recyclable municipally. Categorized as resin #7, or "other" plastics, bioplastics are largely considered a contaminant to recyclable plastics streams. When tainted, these bales go straight to landfill. Because many of the plastics derived from renewable feedstocks are nearly indistinguishable from the petroleum-based plastics they aim to replace, consumers often mistakenly place them in their bins, which causes problems at MRFs.

Some specialized recycling facilities[11] have developed advanced sorting and recycling technology to recycle *poly-lactic acid* (**PLA**, a transparent plastic produced from corn or dextrose) bioplastic waste back into an advanced formula, like

whity2j/Shutterstock

7.4 Does it make environmental—or economic—sense to pour energy and resources into a bioplastic that will be used only once and then thrown away?

E-PLA, that manufacturers can purchase, but these services are cost-prohibitive.

When Are Bioplastics a Sustainable Material?

Durable bioplastics that fit into our existing infrastructure may offer a long-term sustainable solution that moves us away from a dependence on oil. Where conventional plastics such as **PE** (polyethylene, or #4 plastics) and **PET** (polyethylene terephthalate, or #1 plastics) can be recycled, their biobased counterparts (such as *bio-PE* and *bio-PET*) can be recycled with them.[12]

For instance, the PlantBottle, produced by Coca-Cola, is a durable bioplastic alternative to traditional PET beverage bottles. It was launched with up to 30 percent ethanol sourced from plant material and could be recycled with traditional PET containers and bottles. A 100 percent biobased version[13] was recently unveiled.

Combining the recyclability of oil-based plastic with material derived from plants keeps valuable energy and material inputs in the production cycle longer. This makes far more sense than attempting to build an entirely new bio-plastic recycling infrastructure from scratch. Optimistically, the cost of some next-generation bioplastics has dropped to become competitive with even virgin petroleum polymers as biorefineries develop with more and more investments from key stakeholders such as Bill Gates.[14]

Rethinking all aspects of the plastics supply chain in terms of a full life cycle, from sourcing to end of life, is the key to designing a circular plastics economy. While the technology exists to turn things like fruit juice waste, sewage, algae, pine trees, and straw into bioplastics (as opposed to food stock, like corn or potatoes), the infrastructure isn't there for com-mercial biobased polymers to be made on a global scale. Cost is, of course, the determining factor now, but perhaps we can consider some changes to internalize some of these to pioneer the market for sustainable bioplastics.

There is clearly a demand for authentically "eco-friendly" plastics. Leverage your use of more-sustainable, durable bio-plastic by communicating the differences between your efforts versus those of your competitors. Creating value for consumers in this way will offset the up-front costs of a more expensive but sustainable material. Easing the financial risks of wading into the pool of plant-based plastics, governments can subsidize research and farming for the development of new materials.

In the end, for the sector to have the desired effect in a circular economy, consumers need to buy and recycle bioplastic packaging properly. Communicating transparently and diligently about bioplastics will attract conscious consumers, who have a nose for greenwashing and crave authentic ways to reduce their consumption impacts.

Compost may have a hard time jibing with complex materials like plastic, bio- or not. But nurturing resources within a closed system—the planet—can be a no-brainer if all parties know what to do with them.

CHAPTER 8

Less Isn't Always More

Chris Daly
Vice President, Environmental Sustainability,
Europe and Sub-Saharan Africa, PepsiCo

AS PROMISING AS THE SEARCH FOR ALTERNATIVE PACK-aging materials may be, it's a device to sustain continued demand for **consumer packaged goods (CPG)**, which shows no sign of slowing down. **SEE 8.1** They are also known as *fast-moving consumer goods* (**FMCG**). The current global market (valued at $8 trillion) is expected to nearly double to $14 trillion by 2025.[1]

Gts/Shutterstock

8.1 The demand for consumer packaged goods, also known as fast-moving consumer goods, shows no sign of slowing down.

This projection is based on the segment's tremendous expansion over the past half century, facilitated in no small part by the complete overhaul in the way these products are packaged.

Products in the food-and-beverage, personal care, wellness, and household industries have been transformed by the practice of *lightweighting* packaging. This trend shows itself in two ways:

- **Replacing conventional packaging material with a lighter-weight alternative.** A widespread example is the replacement of glass with plastic.

- **Cutting down on the amount of material used in packaging.** This is done by using thinner layers. For example, compared with the 1980s, today's cans and PET bottles use about 30 percent less material.

Lightweighting: An Overview

Lightweighting provides increased access to goods, as the packaging lowers price tags for consumers as well as production and transportation costs for manufacturers. Plastic currently takes up the largest material share in the CPG packaging industry, with a 37 percent share of the total; glass, paper, and metals make up 11 percent, 34 percent, and 6 percent, respectively. Plastic, a comparatively less expensive material, costs less to make and ship and takes up less space in transport.

Increased global mobility, a rise in spendable incomes, and other geopolitical changes, such as urbanization of developing and impoverished regions, have led to a faster-moving world. Many of today's lightweighted packages facilitate the function of convenience items. We see this with food ***pouches*** that can be microwaved or boiled for meals that are ready to serve, that have spouts and caps for easy pouring, and that can be resealed to preserve the product inside.

We also see this with the use of individual portion packs, or ***sachets,*** small, single-use, plastic pouch-like items that are inexpensive to make. They are popular for food, household goods, and personal care products and are used to package everything from condiments and soup starters to shampoo and instant drinks. In industrialized regions they come with our takeout food, as cosmetics samples, and as an easy way to take a vitamin cocktail. **SEE 8.2**

TerraCycle

8.2 Sachets are small, single-use plastic pouch-like items that are inexpensive to make.

Designed for a single use, many lightweighted packaging items are touted as disposable and designed to be thrown away, eliminating the need to clean, store, or otherwise care for an item. Bottled water is a widespread example of this, and most food packaging, shrink wraps, and cosmetics packaging also fall into this category. This is significant for the convenience aspect because hygiene, cleanliness, and "newness" are of value to consumers, as is the turnkey nature of disposable items.

The evolution of packaging has resulted in lower costs and more convenience, which in turn has contributed to the success of a profitable linear production and "one-way" consumption system. We now see that this is creating massive waste problems. Much modern packaging, because of its multi-component complexity or reduced recyclable value, is not recyclable municipally. This is because recyclers also view it as low value, and the cost to recycle these complex plastics doesn't make economic sense. Simply speaking, the cost to collect and process them is more than the recovered value. In combination with the lack of meaningful reclamation systems for the single-use material, packaging just continues to pile up as waste.

This one-way, disposable packaging system is not sustainable. The economic factors that have contributed to the success of the linear system are real and significant, but the waste stream that is emerging (and the speed, size, and variety of that stream) is an issue that needs a radical rethink.

Less Weight, More Waste

The consequence of packaging evolution is that the rate of recycling decreases with each step in that evolution. Glass, for example, though heavy and less convenient than aluminum and plastic, has high value in the market for recyclable materials and is recovered for recycling at a significant rate. Plastic bottles and containers, despite being somewhat recyclable, are viewed as disposable and end up in the garbage. Multi-compositional carton technologies are municipally recyclable in only some regions in the United States,[2] and pouches, sachets, and their ilk are not recyclable at all. **SEE 8.3**

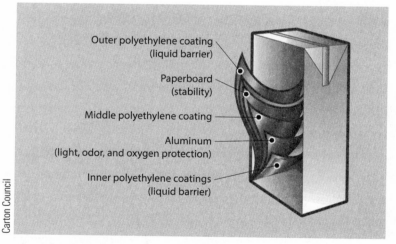

Outer polyethylene coating (liquid barrier)

Paperboard (stability)

Middle polyethylene coating

Aluminum (light, odor, and oxygen protection)

Inner polyethylene coatings (liquid barrier)

Carton Council

8.3 Although they are lightweight and help keep perishables fresh longer, flexible plastic pouches and shelf-stable aseptic cartons are considered difficult to recycle because they are multi-compositional—comprising several layers of different types of plastics and metal foils.

In many cases, a lightweighted item, such as a juice pouch, is **multi-compositional**—composed of several layers of different types of plastics and metal foils. These types of configurations create moisture barriers and protection from UV light, useful in food-and-beverage packaging, but they are difficult to recycle because the components require separating at the material level.

The fitments and closures that give plastic pouches and sacks high function (such as straws, caps, pouring spouts, and spoons) are also not recyclable municipally due to their small size. The loose add-ons fall through the **OCC screens** at municipal recycling facilities and are missed for recovery—or they just get tossed in the trash along with the packages they're attached to, as personnel at MRFs rarely take the time to pull these items apart.

Sachets are prone to ending up in waterways due to their light weight and buoyancy. The challenges are of particular difficulty in the Southeast Asian countries, where they are very popular, as the issue is compounded by inefficient waste management infrastructures. Lids, tear-offs, cracker and chip bags, and the plastic films of shrink wraps have many of the same issues, polluting natural environments and being mistaken for food by wildlife.

The companies that profit from the use of these packaging techniques currently do not cover the costs of designing for their end of life, or the negative **externalities** of waste and pollution that occur as a result. This is lightweighting's biggest problem: no economic recycling model has yet emerged due to the technical challenges in processing and recovering the base materials.

Maximizing Value for Lightweighted Packaging

Lightweighted packaging technologies do not exist in a vacuum. Single-use, disposable packaging wasn't invented because manufacturers sought to pollute the earth. These items drive sales and profits because they provide solutions to real problems and, in many ways, make life easier for consumers.

Giving consumers what they want with a light, convenient package that drives value and is sustainable *is* possible. One way to do this is to choose a lighter-weight material that is highly recyclable. Uncoated paper has been largely replaced by plastic in food packaging. Going back to raw, brown untreated paper for a stand-up granola pouch or bag of coffee beans can create value with a tactile, low-weight package that has a rustic, sensory appeal.

Brown paper bags have pulled their weight for grocery chains such as Whole Foods Market and Trader Joe's, as well as high-end retailer Bloomingdale's. More retailers can do the same, using brown paper for bagging produce, as well as for packing up and wrapping per-weight items such as meat and fish, cold cuts, and bakery bread. Communicating the advantages is key to competing with plastic, to which consumers are now accustomed. An in-store chalkboard listing the benefits of brown paper versus plastic, such as the ability to easily recycle it and fewer chemicals touching the food, is one low-cost way to do so.

Packaging consumables in small product units—one way to lightweight—can be done with durable materials such as glass or stainless steel. Again, consumers are used to a price point and level of convenience offered by less recyclable materials, so driving value is critical. Kjaer Weis, a clean beauty brand, formulates products in smart, weighted silver compacts—valuable, far-from-disposable items that are easy to refill. A high-end brand to begin with, the company has done well to establish itself as a leader in both the premium beauty and zero-waste spaces.[3]

This takes us to the concept of *refillables*, which are a bold alternative for many products currently sold in lightweighted, disposable packaging. Like buying in bulk, using containers that consumers or retailers provide, refilling is very much contingent on retailers willing to integrate the service, as refilling requires infrastructure (i.e., weighing systems, procedures for quality control) to ease use for buyers and to prevent shrinkage. Retailers today sell refills for everything from filtered water and dental floss to spices, craft beer, body oils, and home cleaning products. **SEE 8.4**

Harking back to circular systems of reuse, refilling does have promise. But the single-use, one-way method is what allows consumers to buy what they want, when they want, without the work of taking care of a container or remembering to bring it to be refilled. Because these habits will be slow to change, we must continue to focus on improving the packaging that consumers take home and planning better for what happens to it.

Mattil/Shutterstock

8.4 Refilling and buying in bulk does offer a solution for reducing packaging waste, but it requires heavy lifting on the part of both retailers and consumers.

Reusable beverage pods for single-serving brewing systems are refillable, but they are a durable version of a lightweighting innovation that consumers are less likely to throw away. The ones made entirely of stainless steel, versus combinations of mesh and food-grade plastics, are considered somewhat recyclable, though they may be too small for some OCC screens. Durable beverage pods, many of which are manufactured by companies that produce their disposable counterparts, market their competitiveness with disposable plastic items.

Plastic bottles and containers, the most ubiquitous examples of lightweighting, can be made more recyclable and inspire reuse through design. PepsiCo's premium water brand

LIFEWTR features an oblong bottle shape that stands out from other, stouter, bottle-necked brands on the market. Its labels feature bold, colorful graphic designs by emerging artists that rotate in series of three several times a year.[4] Consumers have been known to refill and reuse the bottle. The striking designs and unique shape give the bottle enough value that a person thinks twice about trashing it and holds on to it to reuse or recycle.

Redesigning packages that have evolved to a form that is no longer easily recycled back into the high-value configurations from whence they came is already happening. With the increasing popularity of premium foods and beverages packaged in glass, such as organic teas and soups, as well as kombucha and other fermented items, we see that consumers are willing to pay a premium for a product presented in a premium package amid a sea of plastic. From a waste perspective, this is a very positive trend.

Glass, heavy and intrinsically high value, is highly recycled but less used today because it is costly and prone to breakage. Plastic has been the mainstream alternative to glass, but other options may exist. For example, packaging in cans what is typically sold in glass is a novel lightweighting tactic that also communicates value through packaging innovation. Many wine and alcoholic cocktail producers have started packaging in cans instead of glass, and from a waste and recycling perspective this is positive when compared with plastic.

Of course, wines are also packaged in cartons, as well as in boxes with unrecyclable plastic bags inside, and these are difficult to recycle. With these and any other unrecyclable pack-ages, recycling options must be improved. Creating, funding, and supporting packaging reclamation systems is essential and is the responsibility of both manufacturers and governments. Producers support such developments through EPR schemes, and these can be much more efficient in driving change.

Small Things Need Real Solutions

For lightweighted items that cannot be recycled through con-ventional channels, the only solution currently is through supplementary *take-back programs* and reclamation systems, a responsibility that producers often strive to shirk.

Take-back programs and recycling platforms for items not accepted municipally do exist and should be expanded. For context, some collections, like the ones for *e-waste* and plastic bags, are often not so much a recycling effort as an attempt to reduce contamination of municipal solid waste streams and ensure proper disposal. Depending on the municipality, drop-off programs are funded by businesses and industry groups, sometimes in partnership with retailers or the local government, hosted at retail locations, municipal buildings, and schools. **SEE 8.5**

Some of these recycling or disposal programs involve consumers mailing their waste to a centralized location. The companies that administer these programs may do so with a fee associated, which is many times cost-prohibitive. But as we see with *regulated waste* (light bulbs, batteries) management services[5] and an emerging market for *turnkey recycling boxes*[6]

Jean Faucett/Shutterstock

8.5 Businesses and governments sometimes administer recycling drop-off locations for items that are not accepted through the municipal program.

for packaging, there are consumers who care enough to step up to minimize their consumption impacts where the public system falls short.

Because accessibility to recycling programs often comes down to cost, a free, easy solution is key. Brands and retailers today do administer take-back programs that are free to consumers for packaging in a number of waste streams. In the face of increasing demand for more ***corporate social responsibility*** and environmental-friendliness, however, the authenticity of some of these recycling programs is questionable.[7]

Take-back programs and in-store recycling promotions are good for business: they generate foot traffic and inspire brand affinity. Hypothetically, a company can incentivize sending in several of its products at a time with the reward of a free item, accepting all brands of its particular waste stream (say, cosmetics packaging) for collection at participating retail locations, and not actually recycle any of it. At little cost to the company, this "greenwashing" engages consumers around a sustainable activity that the company may not even be funding to execute.

Consumers should be vigilant for superficial sustainability claims and, to be sure, ask those companies directly several important questions:

- What are they doing with the materials?

- When are they planning to share a report on the results and outcomes of the program?

- Where is the discarded packaging being processed?

- How is it being recycled?

Companies would do well not to greenwash their packaging and instead focus their efforts on actually collecting and recycling it. Despite not having the internal infrastructure or resources to build a recycling program, conscious manufacturers and brands can still provide consumers authentic solutions at low risk. And consumers want to be able to reward the companies that are doing *real* work.

The challenge with any recycling program, as we know, is cost, so figuring your budget for this type of program will determine how you implement it and in how large a market. The more items there are to recycle, the more resources it will take. Then, look up external vendors you can use to outsource the logistics, labor, and actual processing of the material.

Brands big and small have partnered with TerraCycle, a recycling company that they pay to manage the collection and processing of their post-consumer packaging waste. By outsourcing these services, there is add-on management, infrastructure, and promotion for solutions for companies that may not yet have them internally or that simply wish to leave it to a specialist. The sponsored platform, free to consumers, is often combined with the drop-off format, where retailers, schools, and individuals sign up to collect for their community. **SEE 8.6**

If companies are going to continue producing the packages consumers demand, solutions to the waste problem come

Ashton Brellenthin/Hendricks County Flyer

8.6 Pet food brand Earthborn Holistic partners with TerraCycle to offer retailers the opportunity to host in-store recycling collection as a drop-off for its flexible plastic packaging.

down to recovery systems. Internal or external, reclamation programs cost money, which goes to logistics, personnel, and other operating spends. But the negative externalities of waste must be absorbed by someone besides consumers and the earth. Because consumers are now accustomed to packaging designs that offer high function, low cost, and innovative product experiences, compromises must be made so that the value of packaging compels its return to the value chain.

Smaller Packaging, Bigger Market

Packaging manufacturers, which supply the consumer goods industry with the packaging that is so critical in the production, distribution, and marketing of CPG products, are set for continued profitability as the global population and income levels grow. Analysts forecast the market value for the global packaging industry at $839 billion in 2015, projecting year-on-year growth to reach $998 million in 2020.[8]

Examining the activities that led us to the current, wasteful packaging economy may help us apply solutions that create value where the issues of lightweighting have driven growth. And it's very important that we do because the way the system is set up now, more sales just generate more consumer waste.

For most of the twentieth century, the beverage industry operated with returnable glass bottles ("the milkman")—a *reuse* system, which is one step better than recycling, from a circular-economy point of view. In developed markets the growth of suburbs led to the rise of large grocery stores. The handling of bottles through deposit-and-return systems (setups governed by the *bottle bills* of today) that had been a social tradition at small mom-and-pop stores became an unwanted activity in large-format retail and less convenient for consumers.

Retailers sought to eliminate this activity and make consumption of bottled goods easier. In 1964 Pepsi introduced one-way glass, which was thinner and lighter than reusable glass; it could not be reused, but it could still be recycled. Other CPG companies followed suit, marking the first real shift away from the circular systems of reuse to the one-way disposable system that we know today.

By 1973 plastic in the form of the *PET* (#1 plastics) beverage bottle arrived in the market. Within the space of decade, except for beer and wine, PET replaced glass as the standard package in the beverage industry. This was the first step in a packaging evolution that over the next 40 years would reinforce the linear, disposable packaging model.

With the transition from heavier glass to lighter-weight PET, bottlers could ship more and avoid the risk of breakage. In the 1980s both Pepsi and Coke used this to their advantage, knowing that if there was more product in a home, families would consume more. PET bottles offered this opportunity, and as the standard take-home bottle size grew by 50 percent, beverage sales soared.

The second lightweighting activity was reducing the amount of raw materials used to make the package while preserving its functionality. PET was initially more expensive than glass, but with improvements in plastic technologies over time, the plastic content per package was reduced. Reduced plastic meant less waste when the package was thrown out. This perspective meant that disposal, as opposed to recycling, was less of a material value loss and less of a consequence to manufacturers and producers. The average weight of both plastic bottles and cans has been reduced by 35 percent since the 1980s.

The third and last activity was driven by the ambition of CPG companies to make their products affordable to all. By leveraging new packaging types—including multilayer and multi-compositional materials like juice pouches and sachets—producers could offer their products in less expensive packaging or reduced package sizes at more affordable prices. This made products available to lower-income consumers, extending the reach of the brands.

Convenience Is Currency

All the while this lightweighting evolution developed alongside innovative technologies that have been disruptive and fun, moving with changing lifestyles and consumer habits. More consumers are choosing pouches over traditional glass, paper, metal, and rigid plastic packaging as global market demand was projected to hit $37.3 billion in 2017.[9] Food is the largest and most developed market for pouch use due in great part to rising *output* and consumption rates worldwide. Pharmaceuticals/medical and beverage are the second- and third-largest markets, respectively.[10]

Advancements in seal and barrier technologies for the pouches market are keeping food fresher longer, contributing to a longer shelf life and a greater variety of foods available to consumers. For example, the dairy market segment, which includes yogurt (a product very much in demand), is expected to grow significantly through 2020 with the aid of these high-barrier, aseptic pouches.

In addition to these "wet" goods, which benefit greatly from these seal and barrier pouch technologies, dry goods have also been packaged for convenience. Loose food items such as cookies, nuts, candies, and health supplements are now packaged in stand-up pouches that sit well on shelves or hang from clips in the store, where they had been previously packaged in jars, tins, or envelopes.

As consumers demand quality and ease of use on the go, they also want it at home and in the office. Capsule and pod technologies condense and simplify the preparation of fresh food and beverages and are now common for single-serve, hot-beverage brewing systems, with commercial models for offices, hospitality, and even food service. **SEE 8.7**

The global market for coffee pods and capsules is forecast to grow 9.55 percent between 2017 and 2021.[11] Innovative reclamation systems that piggyback on delivery logistics, collecting from the mailbox where new product is dropped off, turn recycling these items into a premium service for some capsule brands.[12]

The pod and capsule technologies that so successfully disrupted the hot-beverage industry for the greater part of a decade have since been applied to a burgeoning number of food and drink categories, turning food-and-beverage into the most profitable business since software.[13]

Nonfood products are also packaged with thinner, lighter materials that protect the item inside. Electronics, textiles, auto parts, toys, and single-use items like disposable cutlery

8.7 The market for coffee capsules and other pod-based technologies for the food-and-beverage industry is only projected to grow.

and straws are often shrink-wrapped with plastic films or a plastic bag.

Environmental Savings

These single-use strategies have had both positive and negative impacts. Each has offered valuable consumer benefits and has been marketed as being more affordable and convenient and having a smaller environmental impact by taking up less volume. The pod and capsule technologies in particular have been touted as generating less food waste by using the single-serving configuration. Combinations of materials allow perishable foods to sit on store shelves longer, and closures allow these items to continue sitting on shelves in cupboards and refrigerators at home.

The environmental implications of pouches in food packaging and other markets are significant. Pouches are smaller and thinner than glass, paper, and metal packaging; use 60 percent less plastic; and are 23 percent lighter compared with traditional rigid packaging on average.[14] Both the stand-up and flat varieties of pouches generally have a higher product-to-package ratio than rigid packaging and require about half the energy to produce.

A lighter package cuts down on the CO_2 emissions released during production and during transport. Taking up less space means fewer trucks are needed, reducing fuel consumption and additional CO_2 emissions. These are benefits, but of course the downside is a low-value, typically unrecyclable package.

The global population and global incomes will only continue to rise, and collectively we will generate more waste. Significant packaging design and system innovations are needed to drive circular packaging in developed markets. Meanwhile emerging markets must address their infrastructure needs. Waste can be eliminated through a combination of a widening of the compulsory return model (bottle bills), manufacturer- and retailer-administered take-back programs for difficult-to-recycle materials, and the favoring of highly recyclable materials in lighter packaging formats.

Current initiatives aim at reducing or mitigating the negative impacts of the one-way model, but we also need to challenge that model itself. The reasons why consumers migrated to the one-way system, starting around 70 years ago, are today more addressable with new technologies, which producers are in a position to support.

These are all ways to fully close the leaks in the recycling loop, which today remains two-thirds open. Manufacturer investments of time and resources on the design of packages and the systems they flow through will add to the value of your lightweighted package.

CHAPTER 9

But More Isn't Always Better

Lisa McTigue Pierce
Executive Editor, *Packaging Digest*

GOLDILOCKS WAS ABLE TO FIND WHAT WAS "JUST RIGHT" through trial and error, but packaging professionals don't always have the luxury of using costly and time-consuming processes to create the package that is just right for their products. In chapter 8 we discussed the downsides of *too little* packaging, but what does *too much* packaging look like? Sometimes, it's not so obvious.

There is often a good reason behind why a product appears to be overpackaged. For example, by law many health-care products need to tell consumers a lot of things, like how to safely use them and in a certain-sized font that is easily read, limiting how small a bottle or box can be. Other times excessive packaging is supereasy to spot, especially when it seems to serve no direct purpose other than making the product look big or valuable. Ever order something small online—a kitchen utensil, for example—and have it arrive all by itself in a huge box? Ever wonder why it takes so much effort to separate a toy from its package?

While too much packaging can be an issue with any product, it is rampant for upscale products bent on expressing just how special they are so that they can command a high price. This phenomenon is what Tom Szaky calls the "Quagmire of Premium,"[1] which he describes as the trend of premium products using upscale, elaborate, and excessive packaging that is often unrecyclable.

When high-end consumer electronics glue foam to the paperboard carton to cushion the product for distribution and display, the carton is no longer acceptable for recycling. Many perfume sprayers are crimped onto thick, intricately formed glass bottles, making the closures impossible to remove and the glass impossible to recover, despite the fact that glass can be almost infinitely recycled and reused. Some meal kits[2] pack every portion of the ingredients in metallized pouches, comprising combinations of plastic, foil, and paper that are unrecyclable due to their multilayer makeup.

This "Quagmire of Premium" situation is often the most preventable source of packaging material waste. Like an elegant string of pearls, simple packaging can convey high quality if designed properly. How can packages exude value for the consumer without going overboard with frills and embellishments?

Here are seven tips to help designers create premium, attractive packages that don't use excessive or nonsustainable materials—along with examples of brands that got it "just right."

Consider Your Product-to-Package Ratio

What is the correct product-to-package ratio? Well, it doesn't really exist—but that's a great question to ask because the product-to-package ratio can be a critical measure of a product's sustainability.[3] Intuitively, consumers often sense when the ratio is out of whack, as they do with overprotective clamshells, excessive e-commerce delivery configurations, and bulky big-lot bundles.

In the twentieth-anniversary edition of "A Study of Packaging Efficiency as It Relates to Waste Prevention,"[4] the editors of *The ULS Report* identified five ways to prevent packaging waste.[5] One deduction in the 60-page report is that "Ultimately, packaging decisions are driven by consumer perceptions and lifestyle requirements. In many cases, these factors lead to more packaging, rather than less." Why? Because "we tend to equate quality with quantity."

One way that packaging designers can overcome this consumer mind-set is to design a new product and package that makes direct comparison with competitors difficult. Take Tide Pods. The product—which is detergent, stain remover, and fabric brightener—replaces detergent and bleach, each of which previously came in a separate package. Additionally, the product in the pod is highly concentrated, saving more space and requiring less packaging material to contain it. **SEE 9.1**

Granted, this kind of innovation isn't easy, which makes it rare. But, thankfully, it's not the only way to go. Computer tools make it easier than ever to virtually test packaging iterations and find the best design in a short time. Software

Procter & Gamble

9.1 Tide Pods are highly concentrated and dissolve in the wash, saving more space and requiring less packaging material to contain.

from such companies as Stress Engineering Services,[6] Esko,[7] and SpecPage[8] center on primary, secondary, tertiary, and point-of-purchase displays, while a Google search for "pallet optimization software" pulls up thousands of listings, some of which deal specifically with optimizing packages on a pallet.

Reduce the Number of Packaging Components

How many packaging pieces or layers are truly needed? Take a cue from nature, which knows how to protect itself—and pretty efficiently. How much more protection does a coconut need? The sturdy husk keeps the product fresh and pure until it's cracked open. Process that into coconut water, a drink

Genuine Coconut

9.2 Genuine Coconut Water takes premium to the next level with one layer of packaging that communicates authenticity and purity: its own hard shell.

popular with today's health-conscious consumers, and you need a bottle, a screw-on cap with a tamper-evident ring or an inner seal, and a label.

Or do you? Genuine Coconut Water decided to simplify and take advantage of the package Mother Nature had already designed.[9] Minimally packaged, the coconut's patented easy-opening system is a ring made of recycled coconut husk fiber and natural resin; consumers pop it open like they would a soda can. This allows the brand to tout that Genuine Coconut Water is a 100 percent organic certified, authentic, ecological, and almost totally biodegradable beverage because it uses only natural coconut resources. **SEE 9.2**

Gafs Kartong/Highcon

9.3 Carton design and production by Gafs Kartong in Sweden, a Highcon customer, add texture and visual appeal without rendering the package unrecyclable.

Many luxury products—liqueurs, antiaging skin creams, chocolate truffles—add shiny foil to their secondary carton. These shimmering packs are effectively eye-catching but are not easily recyclable. If the bottle, jar, or tray must be put in a carton, is a monomaterial a possibility? Intricate die cuts[10] made by digital lasers might just be the ticket. The attractive, nonfoil carton can then be recycled. **SEE 9.3**

Debossing or embossing also adds a visual element that elicits a tactile response—inviting consumers to pick up the product—all without interfering with recycling. While typically seen on paper packages, this effect is now also available on highly recyclable aluminum. The uShape technology[11] from Montebello Packaging enables asymmetrical shapes for

Montebello Packaging

9.4 Montebello Packaging's uShape technology bottles are both recyclable and beautiful—the best of both worlds.

aluminum beverage bottles, such as flutes, embossing, debossing, and various fine details. **SEE 9.4**

Design for Recyclability

If your package can't be made from just one (preferably recyclable) material, the next best option might be to make a package with multiple materials that are easy to separate. That's a design feature of one toy package that earns high marks for creativity.

For its Moana doll, Disney designed a package that allows consumers to separate the plastic window from the carton with instructions and messaging that encourages proper recycling of the box's components. It also encourages kids to use the rest of the Moana doll packaging in their play.[12] With a few folds, the inner packaging transforms into a boat, which factors strongly into the movie's plot of exploration and discovery. **SEE 9.5**

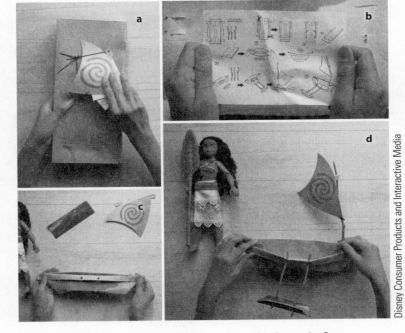

Disney Consumer Products and Interactive Media

9.5 Disney's Moana doll packaging earns high marks for creativity. **a** The inner carton is separate from the outer shell. **b** Simple instructions encourage children to use the packaging in their play. **c** With a few folds, the components transform into an outrigger sailboat. **d** The packaging itself becomes a part of the child's imaginary adventure.

Avoid Obsolete Graphics

Change happens incredibly fast these days. New recipes, updated regulations, seasonal specials, limited editions, personalized labels—brands constantly vary package graphics. The trick is not to accumulate an inventory of obsolete designs that must be trashed.

Digital printing[13] has evolved exponentially in the past decade, giving brand owners flexibility they never had for switching up graphics instantly at a quality often comparable to preprinted packaging materials. But there's another "graphics" technology that you might consider, especially if you're looking to expand your packaging for today's "connected consumers": *augmented reality (AR)*.[14]

A leading US producer of pasture-raised eggs, Vital Farms, updated the design for its egg cartons to leverage the interactive power of smartphone-enabled AR.[15] The chalkboard-style graphics on recyclable paperboard already convey a farm-to-table message that speaks to the company's philosophy and farming practices, while the AR component shows, rather than tells, consumers exactly what they are buying, adding an element of fun that makes for a memorable consumer experience. **SEE 9.6**

Change the Channel

Packaging for items sold in retail stores must protect the product, from manufacturing to its point of use. It must also catch the consumer's eye, conform to the retailer's available shelf space, and compete effectively side-by-side against its opponents.

Vertical text on right side: ROAR Augmented Reality

9.6 Vital Farms uses augmented reality to drive home their farm-to-table message using interactive technology.

But a lot of those requirements disappear when the product is sold through e-commerce. Online stores don't have shelves, so package size is no longer an issue. This allows brands to rethink the optimum packaging for the use of the product,[16] freeing designs from shelf height or having to maintain a visual size comparison with competitive products.

And because the product, rather than the package, is more likely displayed, expensive "romance" packaging is not needed, says Brent Nelson, senior manager of packaging and sustainability at Amazon.[17] "Packaging designed to stand out on a retail shelf is often oversized, with...redundant features to prevent theft and [is] not capable of surviving the journey to the customer. In many cases, these features can lead to sub-optimal packaging for online distribution."

What packaging is optimal for small-parcel, direct-to-consumer delivery? Amazon actively works with brand owners to choose packaging that satisfies all parties.[18] "Certified packaging designed for Amazon and online fulfillment is a win for customers due to right-sized packages being designed to prevent damages; it's a win for our brand owner partners and Amazon because it's less material volume and often much lower cost; and it's a win for the planet."[19]

One Amazon Frustration-Free Packaging case study is the Norelco One Blade shaver project. By rethinking the packaging that was critical for protection during shipping, Norelco brand owner Philips cut the number of components from 13 to nine and realized an 80 percent reduction in packaging volume.[20] **SEE 9.7**

There is still room for improvement for e-commerce packaging from a sustainability point of view. As Bob Lilienfeld, co-author of the whitepaper "Optimizing Packaging for an E-commerce World,"[21] says, "We can't look at the concept of sustainability in a vacuum. We must look at the entire system of product containment, protection, storage, delivery and end-of-life."[22]

Think Holistically

Most sustainability-minded people know to look at the big picture because sometimes a change in one area causes unintended but disastrous consequences elsewhere. But sometimes, happily, the opposite happens. A small change in the shipping

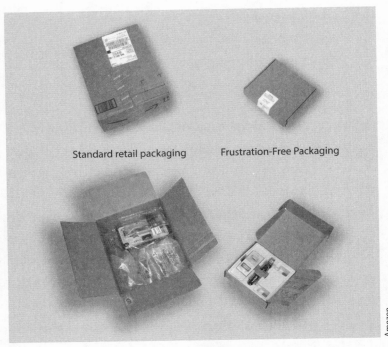

Standard retail packaging   Frustration-Free Packaging

Amazon

9.7 Amazon's Frustration-Free Packaging certification program works with brands such as Philips Norelco to reduce packaging volume while satisfying consumers.

package might allow huge material reductions at the primary pack level. Or a dramatic change in the primary pack could cause significant savings throughout the supply chain.

"Waste means a lot of different things. The most obvious is trash," says Jason Foster, founder and CEO of Replenish. "But it's also wasteful in time, resources, money....In light of this, there's a better way to design the products that we use every day to be less wasteful."[23]

Foster invented the refillable Replenish bottle system with a holistic view of product/package value.[24] Designed to

Replenish

9.8 Replenish's customized refillable bottles system offers concentrated product and a solution that wastes less time, resources, and money.

be durable and withstand years of reuse, a pod of concentrated liquid (different iterations include cleaners, beverages, and soaps) connects to the bottom of the bottle and provides up to six refills. After buying one Replenish bottle, consumers simply purchase the smaller pods, saving about 80 to 90 percent of plastic, energy, and pollution. **SEE 9.8**

And that brings us to one last piece of advice for how you can optimize packaging.

Choose Partners Wisely

No company operates in a vacuum. Unless you are designing, manufacturing, and packing your products in-house with technology and production lines you own, you are buying these supplies and paying for packaging services from external vendors. Partnering with innovative packaging suppliers also committed to minimizing the impact on the environment enables positive collaboration along the production matrix.

Here is an example: The first packaging company to achieve the *Cradle to Cradle Certified* (*C2C*) product standard was Be Green Packaging,[25] whose line of 100 percent molded plant fiber supplies undergoes an annual life-cycle assessment for production, factory operations, and end of life. Offering compostable, "soil-to-soil" solutions for the CPG, food service, and industrial packaging industries, Be Green provides access to its waste audits—cultivating transparency—and directs customers to the BioCycle[26] composter portal, ensuring access to organics recycling.

Third-party certifications from reputable organizations are beneficial as a means to give confidence that products and packaging are, in fact, delivering on the performance indicators their designers claim. Used by Whole Foods Market for its iconic "utility trays" and for Procter & Gamble's Gillette Fusion ProGlide Razor, Be Green Packaging is also currently certified by the USDA BioPreferred Program, the Non-GMO Project, and the Biodegradable Products Institute. It stands out in the packaging industry by helping customers improve their impacts. **SEE 9.9**

As concern for our planet stays top of mind for many shoppers, designers can leverage the halo of sustainability by using any or all of these tips to optimize their packages and support

Be Green Packaging

9.9 Gillette's Cradle to Cradle Certified (C2C) Fusion ProGlide Razor package would not have been possible without a smart partnership with Be Green Packaging.

recycling—two of the three new sustainable packaging priorities of megaretailer Walmart,[27] which sparked much of the interest in packaging sustainability[28] with the introduction of its Packaging Scorecard in 2006.[29]

When brands offer packages that connect emotionally with consumers and enhance their experience with the product, they create memorable "aha!" moments that build loyalty and encourage customers to buy again.

CHAPTER 10

The Forgotten Ones: Pre-consumer Waste

Tony Dunnage

Group Director, Manufacturing Sustainability, Unilever

EXCESSIVE PACKAGING FOR PRODUCTS AND E-COMMERCE is something consumers notice and have been quick to criticize. There is a focus on the packaging waste they see and interact with, called *post-consumer waste*. But there's more, as a significant amount of packaging and manufacturing waste is generated *before* a product reaches shelves—and in greater volumes and varieties than people realize.

What Is Pre-consumer Waste?

This waste stream, known as *pre-consumer waste*, is defined as material produced during the manufacturing process that does not make it into the final product. This includes but is not limited to the following:

- Packaging and wrappings used in the transport of raw materials from supplier to manufacturer

Goritza/Shutterstock

10.1 What happens to all the shipping and transport waste that gets products and their packaging from point A to point B (and C and D and…)?

- Trimmings and extras from primary, secondary, and tertiary packaging or product manufacture

- Shipping and pallet waste from the transport of products and packaging from manufacturer to distribution center to retailers **SEE 10.1**

Through this lens one can also consider products and packaging ordered in overstock, damaged, printed with out-of-date artwork, or returned to be pre-consumer waste.

Even though consumers may see it in a retail or advertising capacity, *point-of-sale* (**POS**) material (such as in-store signage, merchandising units, and billboards) and *secondary packaging* (what bulk products come in, like a crate for soda or a case of boxes of cereal) also fall into this category. Pre-consumer waste is rarely visible to the customer, is frequently overlooked during packaging design, and often ends up in the landfill or incinerator due to the costs of reintegrating it into the supply chain.

Business as Usual

Producers and manufacturers should be responsible for the end of life of packaging and what happens to it after consumers are finished. In the case of pre-consumer waste, there is even less room for attempting to push this responsibility downstream, as this material is still within the control of the factory, transport, and retail systems. Producers invest time and money into these systems, so it makes sense that they would retain accountability for their operations, including the management of their surplus and waste material.

With little economic incentive to account for these items and a lack of regulations compelling them to do otherwise, however, producers and manufacturers often remove themselves from responsibility, leaving someone else to deal with their pre-consumer waste. For example, POS and merchandising material has no use to the manufacturer after it serves its purpose of promoting a specific product for a particular season. It's out of date and out of style, so the retailer doesn't want it, either.

Without a reclamation system to keep this material with the manufacturer, the retailer, lacking the infrastructure to handle the material, throws it in the trash. The same goes for the secondary and tertiary packaging—the shipping and pallet materials that manufacturers and distributors use to bring products to market. Once the products are delivered, they no longer consider it their responsibility, leaving it with their retail customers to dispose of as they will.

Wasting such materials does come at a cost, but so do the processes to convert them. When it costs more up front to develop and implement these processes than it does to send the materials down the linear disposal route, off to the landfill or incinerator they go, valued as a negative. The result is often the loss of potentially useful material, supply-chain and resource maximization, and potential revenue streams, along with the political, social, and environmental externalities of waste.

Reverse logistics—the process of diverting usable material from linear disposal for the purpose of reusing it or capturing its value—necessitates systems that initially require resources to develop and execute. The needs and capabilities of the small packaging vendor differ from those of large CPG companies, which have their own factories and supply chains. But the path by which usable materials become waste is largely the same.

Pre-consumer material starts out with some worth in the eyes of the businesses that produce it because they are in possession of it. Manufacturers have long been keen on reusing and repurposing scrap in various ways. Supplying such materials to other companies or reintegrating them back into their own production lines ensures that resources continue to work for manufacturers, their customers, and the planet.

Current Challenges and Trends of Business-Driven Action

Full disclosure: In 2008 Unilever's factories sent 140,000 metric tons of nonhazardous waste to landfill—that's the equivalent of 2 million waste bins, or 17 Eiffel Towers. Given what Unilever has since managed to achieve, it almost seems unreal that we have ever done things any other way. But as is the case for any company today, the first steps on the sustainability journey entailed coming to terms with the fact that the business-as-usual way of doing things was no longer working and needed to change. **SEE 10.2**

In addition to an ambitious target of using 100 percent reusable, recyclable, or compostable plastic packaging by 2025,[1] Unilever set the important, pivotal goal of achieving zero nonhazardous waste to landfill[2] by 2020. *Zero-waste-to-landfill* **(ZWTL)** is a business concept in which solid waste produced at respective facilities is not landfilled but instead is reused, recycled, composted, or disposed of via some other outlet. Some companies regard ZWTL as a guiding ideal rather than a benchmark,[3] and many seek certification[4] for visibility.

At the core of these programs are continuous improvement and constant evaluation about material choices. There is not a one-size-fits-all approach to reducing waste in factories, but a strong commitment to eliminating waste at the source is the preferred place to start. Optimizing manufacturing to prevent trimmings from production prevents the need for reverse logistics, the next preferred step for resource management.

10.2 Unilever's factories once sent 140,000 metric tons (or 2 million waste bins, or 17 Eiffel towers) of waste to landfill in one year. Their zero-waste journey started with the realization that "business as usual" had to change.

The fast-moving and consumer packaged goods in the Unilever portfolio, which spans food and beverage, cleaning agents, and personal care products, have their own unique manufacturing streams. But thinking outside one's sector of expertise to brainstorm on other industries may tap the creativity and imagination to come up with novel solutions for the business at hand.

Take textiles, for example. A durable good (all things considered, *fast fashion* notwithstanding) and far from the CPG realm, textiles produce huge waste streams both pre- and post-consumer. Optimized machines producing knitwear and sweaters in one piece, as well as cutting cloth from a pattern that keeps most of the source material intact, reduces the use of material and the need to recycle or dispose of it. The same concepts of optimization and one-piece production can be applied to fast-moving packaged goods.

Unilever successfully applied weight reduction as a production tactic, which has been implemented by many other CPG companies. Less weight for a package equals less scrap potentially sent to landfill. But unlike post-consumer, it's still within your control and not up to municipal systems to recycle. One condiment company reduced the size of its bottles, openings, and spouts. This reduced the number of trimmings from molding the bottle, as well as the amount of wrapping material used to seal it.

The next way to eliminate waste sent to landfill is to integrate the inevitable production scrap. For the same reasons that separation has been the key to successful post-consumer waste diversion, much more action has occurred in pre-consumer

Unilever

10.3 Separation is one of the most important aspects of recycling, and the integral factor in maintaining resource utility for pre-consumer waste materials.

waste reduction. Because the waste streams are comparatively uncontaminated and easier to separate at the point of generation, this is true in both the durable and nondurable goods sectors. **SEE 10.3**

Maintaining separation of materials is integral to this achievement and the expansion of reuse to other activities. One Japanese electronics and technology company developed a way to convert material from its solar panel production plant into usable powder for cement. As a result, the site's factory waste "recycling rate" reached 93.5 percent (it is important to note that "recycling" into energy is the least favorable circular solution, though it's better than landfilling).[5] "Before" and "after" charts prepared and shared by each region have accelerated recycling improvements at underperforming locations.[6]

For Unilever a detailed mapping of every site's unique mixed waste streams maintained material segregation through the provision of dedicated waste collection and storage points; this was the first step to developing action plans for reuse, recycling, or recovery, which were made available to all locations. This approach has meant reconsidering every single material that is consumed and encouraging transparency across operations.

Distance and the unique geographies and regulatory bodies of different locations are always challenges for reverse logistics, which require shipping, transport, and quality control of materials. Unilever has developed a number of replicable solutions and applicable contingencies for tackling waste:

■ A solution developed in Kenya repurposes laminated wrapper waste into a versatile rigid construction material that can be formed into corrugated or flat panels for use in fencing and roofing.

■ A vermicomposting process in South Asia uses worms to break down manufacturing food waste into compost, which is used to fertilize organic vegetables for staff canteens and made available to employees and local communities for use in growing their own crops.

■ In a UK factory, raw and packaging materials are received in cardboard boxes. It has been found that the boxes can be reused by other industries directly without the need for recycling. This is not counted as waste because it helps reduce the consumption of natural resources.

■ In Australia reject stock tubs are turned into fertilizer. And a partnership with an innovative waste management company[7] recycles surplus plastics from the food-and-beverage unit into underground pipe covers and garden edging.

■ Tea bag paper from Russia is recycled into wallpaper.

■ In Africa and Latin America,[8] hard-to-recycle materials are shredded and compressed at high temperatures to make school desks.

These sorts of reverse logistics were facilitated with direct relationships between supplier and manufacturer, strengthened and developed through open dialogues. While collaboration is important, retailers and manufacturers will be more likely to accept and pilot innovative processes when they can see the value and sensibility of investing in them. By developing systems that worked for each facility, and then deploying to figure out how they could work for others, Unilever not only captures the value of material but generates gains through systems creation.

For example, Unilever updated pallet systems from the single-use wood pallet to reusable, durable plastic pallets that get repaired. Standardization of items such as pallets and intermediate bulk containers facilitates multiple uses and return systems that work across the industry—moving from single-use to multiple-use. This drives value for the facility and can be applied for others, creating a larger infrastructure for waste reduction and reuse.

Unilever

10.4 At some manufacturing locations, repurposed laminated wrapper waste is converted into a versatile rigid construction material for products such as recycling bins.

Here we ask: Why can't this be done with other materials? Why not a collapsible plastic box? Why must secondary packaging and distribution and POS material be one-way, the energy and resources poured into them lost forever to linear disposal? By instead putting these items to good use, you can create value and save money for your suppliers, your customers, and other industries. **SEE 10.4**

One company achieves its zero-landfill goal by convincing suppliers to ship parts in reusable containers, which allows packaging for some 80 engine parts to be used multiple times instead of once.[9]

No matter what your manufacturing sector—durable or nondurable, packaged, luxury, or service-based goods—thinking about ways to get the most out of the materials you invest in will unlock possibilities at every turn.

At Unilever sites in many countries, waste material is recovered and used as alternative fuels to generate energy through a global partnership with a cement manufacturer and its waste management service provider. Combustible waste material, which was previously sent to landfill, is used as fuel in the cement manufacturing process. In Sri Lanka waste tea leaves are used as fuel in boilers, which also helps reduce carbon emissions. The material is pretreated and used as alternative fuel and raw material in cement kilns. Even the ash is used—and it is fully incorporated, thereby leaving no residues.

Despite the growing popularity of ZWTL in sustainability circles, many organizations still harbor the mind-set that it does nothing for the bottom line. You will find that it is quite the opposite.

The Impacts of Reducing Pre-consumer Waste

The early pioneers may not have taken the easiest or fastest path, but like the Wright brothers and their first airplane, it takes someone to prove what is possible. The Wright brothers didn't go the farthest, and it wasn't the most comfortable, but it was one of the first man-made machines driven by humans to fly through the air. Setting out to demonstrate that it is possible to reach goals that seem unachievable today is essential. **SEE 10.5**

Ultimately, Unilever achieved the ZWTL goal in 2014—six years ahead of target. This meant that more than 240 factories in 67 countries reused, recycled, or recovered all of their nonhazardous waste and sent nothing to landfill. Unilever's

Library of Congress

10.5 Like with the Wright Brothers and their first airplane, it takes someone to prove that pre-consumer waste diversion is not only possible but replicable.

sustainability leaders believed that the model and mind-set that drove the achievements in these factories would be repeatable beyond a manufacturing environment.

So, they set out to extend this to other parts of the business. By February 2016, nearly 400 additional Unilever sites, including offices, distribution centers, and warehouses in more than 70 countries, had achieved zero waste to landfill. Today Unilever sees no landfill waste in its factories, has proud and inspired employees, has achieved $234 million in annual savings and costs avoided (to reinvest back into the business), and has created 1,000 jobs in the wider economy.

They did all of this while reducing waste at the source and eliminating overconsumption of inbound materials. As a proud member of the *CE100*—a global platform bringing together leading companies, emerging innovators, and regions to accelerate the transition to a circular economy—Unilever employed circular-economy concepts[10] to design renewably and reduce the amount of material used in production. This is the best solution environmentally and where the biggest savings are realized. By 2016 waste per metric ton of production was reduced by 37 percent compared with the 2008 baseline.

By bringing nonhazardous manufacturing waste to landfill to zero, Unilever unlocked two huge benefits. First, at a conservative but accountable estimate, 450 new jobs have been created for disabled and disadvantaged citizens in poorer parts of the world as a direct result of strategies for redirecting by-products and "waste" from manufacturing to new and useful purposes. **SEE 10.6**

Second, groundbreaking momentum for the business was driven with employees around the world. Shortly after the completion of these targets, Unilever held an event with competitors, peers, suppliers, and partners, inviting them to join this important journey.[11] The presentation shared details of the methodology and introduced and highlighted best practices, hoping to amplify them across the industry. By designing once and deploying everywhere, Unilever hopes to help every business find its own model for zero waste.[12]

Sharing the model, rather than keeping it under wraps, is important for transparency—one of Unilever's core values and an aspect of its culture. A transparent account of progress through engaging, insightful, and evidence-based reporting

Unilever

10.6 ZWTL creates jobs and engages employees and stakeholders around the deployment of methods and models across industries.

allows the garnering of the best feedback from consumers and stakeholders to keep improving.

How Do I Do This in My Facility?

The first steps toward ZWTL can be difficult. Begin by focusing on what is most within control: the factories.

1. Start by analyzing facilities and processes you can easily control. Perform waste audits for your locations, testing each individual product, department, and process, to get a comprehensive idea about the factors influencing waste generation for your company. Taking stock of both positive and inefficient material flows will show where there's room for improvement and highlight areas that can serve as models for other processes.

If you are a packaging buyer or small company that uses external vendors, ask to review their latest waste audits to get an idea of how sustainable your product really is. How much waste does your supplier send to landfill? Though your immediate operations may be relatively low impact, those of your vendors may not be and are thus worth a look.

2. Understand the changes and cost-cutting procedures that are possible. Based on the data gathered from a waste audit, potential changes in production processes, materials usage, and changes in the operating procedures governing waste activities will emerge. Perhaps a piece of machinery can use an update. Maybe you can commit or reorganize staff for a particular resource management team.

Mike Robinson, vice president of sustainability and global regulatory affairs for General Motors (GM), told *Forbes*: "A landfill-free program requires investment....It's important to be patient as those upfront costs decrease in time, and recycling revenues will help offset them."[13] When GM started its commitment to landfill-free manufacturing in 2005, it invested about $10 for every 1 ton of waste reduced. Over time it reduced program costs by 92 percent and total waste by 62 percent.[14]

Committing to a goal of diverting waste while considering costs may impact current supplier services, operations, and waste-handling practices. Audit and simulate different scenarios to figure out what is possible for your facility for the current year, in five years, and over the next decade. Cutting costs and improving your sustainability numbers may be a matter of working with different partners, investing in new materials, or eliminating a division.

Whether you outsource a consultant or dedicate the resources to develop your own sustainability division, under-standing your company's needs and priorities is one of the most important aspects of making a case to management, who ultimately must sign off on changes.

3. Get the support of company leadership. Not only do you need to get leadership on board for budgeting and resource purposes but you need to get them engaged in the communi-cations that set the tone for the ZWTL journey. For Unilever, senior leadership support, with regular updates on the topic coming from the chief supply-chain officer, helped inspire and gain traction for ZWTL throughout the organization.

Changes in day-to-day processes and company structure in the name of reducing waste will not work without inspira-tion, motivation, and engagement from top-down leadership. Keeping everyone in the loop with reporting, information, and regular best-practice seminars and meetings is an investment of time in your personnel—the real resource.

4. Keep materials within your control when possible. If you have the capability, keeping materials within the closed sys-tems of your factory locations is ideal. It eliminates the need to prepare or transport materials for sale, which comes at a cost, or to disclose sensitive information to vendors. Utilizing pre-consumer by-products to generate revenue or incur cost savings yields win-win situations.

Mapping waste streams and estimating costs helped Unilever's factory teams understand where costs for disposal could be transformed into revenues through recycling and reuse, further supporting the case for action. Working with suppliers to reuse shipping materials is one of the simplest ways to do this. Materials that go back and forth over the same route are relatively easy to control and don't require too much change from business as usual, save that of training staff and partners to reuse instead of grabbing new.

Having a strong network of suppliers committed to working with you on closed-loop systems can help you recycle your own factory waste into something you'd pay for anyway. For example, one of GM's plants turns paint sludge waste into plastic material for shipping containers durable enough to hold engine components.[15]

5. Find markets for waste material. Commitments to go ZWTL can be hampered in parts of the world where the recycling infrastructure is lacking. As we know, this is most places, so finding markets for pre-consumer material can be a challenge. Working with your partners and suppliers to determine a fit within your network will help address these gaps. Viewing your waste as a marketable commodity is about finding a demand in the market for your supply.

Manufacturers in both your industry and others stand to save money by buying leftover packaging trimmings while allowing you to maximize the value of the material you purchased for production. Companies already exist that buy and sell commodity scraps like leather, molybdenum alloy, and pre-consumer paper; you could be the supplier. There's a decent market for pre-consumer plastic waste, such as industrial packaging, as it is available at scale from fewer sources, making for a desirable product.

Creating a market for your materials by supporting the local recycling system is also important. Invest in new and emerging technologies to best position yourself. Micro-factories in Australia, for example, which show promise as a solution to current, centralized recycling facilities, are able to operate on a site as small as 50 square meters and can be located almost anywhere.[16] Using these and other types of solutions in development may give them the boost they need—and you a nice discount on services as their initiatives continue to grow.

6. Aim for early wins. At the start of the journey, Unilever knew it needed an early win and a repeatable reference to prove success once and then deploy widely. So, leaders decided to first focus on bringing ZWTL to operations in the United Kingdom. Data gathering, proposal tendering, and supplier selection began in 2010, and a nationwide partner to manage manufacturing waste was appointed. When the contract went into effect in July 2011, all 29 of Unilever's UK locations, including 12 factories, achieved **ZNHWTL**, or *zero-nonhazardous-waste-to-landfill*, in a stroke.

Getting there had taken 11 months, but they had executed a template that could be rolled out across Europe. By the end of 2011, they'd successfully repeated the model to zero-waste manufacturing in Belgium, the Netherlands, and Luxembourg. Within a few more months, Italy too had achieved ZNHWTL.

7. Track progress and share resources. The key to recovering the highest value from pre-consumer waste is managing all by-products in one electronic tracking system. All Unilever plants monitor, measure, and centrally report their performance on a monthly basis, and the data are evaluated against companywide waste reduction goals. By engaging employees in the recycling effort, the data also help motivate factories to keep looking for creative solutions.

If one plant finds a valuable use for a by-product, it is quickly shared with other factories around the world. Reporting includes performance data for direct operations, as well as information about managing material issues across the extended value chain, which encompasses both the supply chain and how products reach consumers.

Practical activities such as "dumpster dives," where Unilever factories emptied bins in front of management teams, demonstrated the need for recycling at each location. **SEE 10.7** "Zero Hero" communities across geographies shared practical examples and experiences and collectively addressed issues. Zero Heroes highlighting successes and methods were fundamental in accelerating the program's exponential performance.

Impacts can be far reaching. Identifying and prioritizing the most material impacts to your business and stakeholders to measure quantitative performance indicators will help monitor the development of specific risks over time.

Dr. Bronner's

10.7 Sorting and tracking activities such as dumpster dives take stock of company material streams and reduce the amount of waste sent to landfill, diverting what can be recycled or composted.

Pre-consumer Waste: A Resource Management Issue

As one of Unilever's employees said on the company's ZWTL journey, "It's like working on a mosaic, and every stone is worth adding to make it complete." By setting a clear vision, creating a knowledge network, and empowering your people to act, you can inspire your employees and suppliers with zero-waste ambition.

Getting everyone in your organization and those you work with on the same page will not be easy—but it will be worth it. Full support and commitment from leadership streamlines processes and provides structure from the top down. Meticulous planning to deal with the unknown, total and utter commitment to the task, and a passion to succeed are the guiding elements of designing for zero waste.

Packaging design includes the processes associated with manufacturing, transporting, and distributing it. To make the most informed decisions about materials, suppliers, and reuse-and-recovery routes, producers must examine best practices from the manufacturing process for every link in the supply chain. Pre-consumer material becomes pre-consumer waste when business is unable to implement the systems that capture its value. Preventing such waste and increasing profits starts and swells with a plan.

Business-to-business waste management allows you to control every aspect of material flow, which saves time and money. Investing in reverse logistics—capturing value by moving pre- and post-consumer goods from their typical final destination of the landfill—takes full responsibility for your production.

Remember that sustainable supply chains are about people, not just processes. Human resources, inside your organization and out, are a renewable source of energy that when nurtured yield limitless returns. Stakeholders and consumers crave accountability, transparency, and learnings, and they care about who companies are at their core. By addressing problems with waste at the source, you carve out an essential place in the lives of consumers and the world at large.

Consumers Care

KoAnn Vikoren Skrzyniarz
Founder and CEO, Sustainable Life Media,
producers of Sustainable Brands

O VER THE PAST HALF CENTURY OR SO, OUR COLLECTIVE sense of whether we are living a good life has shifted away from the original values of "life, liberty, and the pursuit of happiness" that shaped our country. Those values that united us as citizens have slowly given way—thanks in no small part to the advertising and marketing industries—to our seeing our collective selves as consumers more than citizens, as little more than commoditized objects to serve the global economic engine.

When the word *consumer* was first used as early as the Middle Ages, it meant "one who squanders or wastes."[1] Today the term can be used to describe each and every one of us. Our modern-day consumer culture—which has become inextricably tied to the growth of economies and the industries that feed into them—has fettered our individual consumption to our measure of financial, social, and personal well-being.

But the pendulum is beginning to swing the other way. A growing contingent of citizens around the world understand that our time—the most precious commodity there is—is controlled by the pursuit of money and material things. More of us are realizing that our overconsumption has negative impacts not only on our own health and that of our communities but also on the health of our planet.

A Role for Business

Sustainable Brands research shows that 89 percent of consumers in the United States believe that if people understood each other better, we would all live a better life.[2] Citizens around the world understand that consumption itself does not equate to happiness and that our obsession with keeping up with the Joneses can actually get in the way of it.

As a result, consumers are slowly but surely reprioritizing their aspirations and, in keeping with this shifting sense of priorities, expecting more from the brands they choose to support. Consumers care about the impact of their purchases on not only their own health and that of their families but also the health of their communities, the environment, and the broader ecosystem of people involved in making the things they buy. We want to know where and how the products we buy are made, how the workers who make them are treated, and where the goods and their packaging will go once they're no longer in use.

Brands stand to gain from aligning with consumers, who increasingly engage in environmental stewardship, social good, and—you guessed it—sustainability. A growing number of hurricanes, wildfires, and extreme weather events shake the entire planet, in a world where news is global and available at the tap of a smartphone. Consumers are irate that microplastics and fibers are showing up in their food and drinking water and that companies continue to profit from a system in which they have no responsibility to the planet or its people.

Getting to the core of what people truly care about is what drives value for business. When it comes to our aspirations, we are more alike than the media makes us out to be. Across generations (34 percent of millennials, 35 percent of Gen Xers, and 38 percent of baby boomers or older—66 percent of the population overall), the interest in ***balanced simplicity***, or living a simpler, healthier life, rises to the top as having the largest impact on defining a good life for today's consumer.[3] **SEE 11.1**

Next in line in terms of impact is the want of ***meaningful connections***—the desire to be closer to family, one's community, and the environment. Both balanced simplicity and meaningful connections top money, status, and personal achievement as aspects that more accurately reflect the life consumers want to achieve. While the media highlights our differences in order to lead us to define ourselves through consumption, there appears to be a shared optimism about the possibility of a meaningful existence.

Enabling the Good Life

Consumers are tired of the negative impacts of their consumption and are ready for a change. People want business to help them live more meaningful, balanced lives, and a strong majority (80 percent) say they are loyal to those that do.

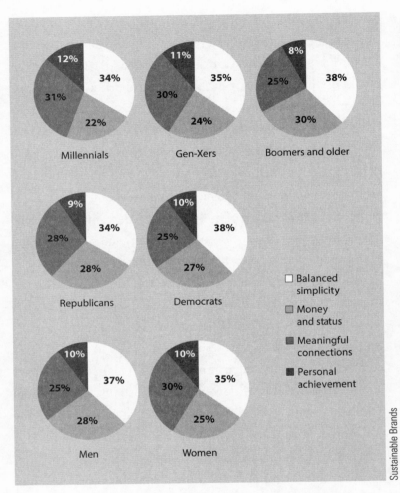

Millennials

Gen-Xers

Boomers and older

Republicans

Democrats

☐ Balanced
   simplicity

☐ Money
   and status

▨ Meaningful
   connections

■ Personal
   achievement

Men

Women

11.1 The interest in balanced simplicity, or the concept of living a simpler, healthier life, defines a "good life" for consumers across generations. *Note: Percentages may not total 100 due to rounding.*

But while 51 percent sense that companies would like to help them live the life they seek, nearly two out of three consumers struggle to name brands that are actually doing something about it.[4] **SEE 11.2**

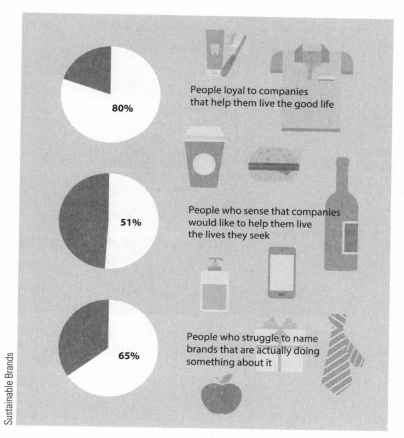

Sustainable Brands

11.2 Brands that offer consumers the opportunity to live the simple, balanced lives they seek stand to gain in an increasingly competitive marketplace.

Taking the time to find out what people really want can help companies establish purposeful missions and sculpt operations around those values. Start-ups can take a page from Lush Cosmetics, a company that makes little to no packaging[5] work for it through value creation and cultivation of brand identity. In addition to reusable metal tins, colorful cloth knot-wraps,

Lush Cosmetics

11.3 Lush Cosmetics stayed true to its start-up vision of recycled and no packaging and grew into an international category leader to which consumers are loyal.

and 100 percent post-consumer recycled plastic bottles and pots[6] (some of it ocean plastic),[7] 35 percent of Lush products (including solid shampoos, conditioners, massage bars, soaps, and bath bombs) are sold "naked," allowing consumers to touch and smell them through a unique retail experience that harks back to the shops of yore. **SEE 11.3**

From a start-up partnership formed in a beauty salon in Poole, England, Lush was able to grow and globalize while maintaining the circular packaging and production practices that gained it a cult following in the first place. "If we don't grow, we won't see the change we need… As with all dreams, it ends in responsibility," says Lush Cofounder Mark Constantine.[8] With ethically sourced ingredients, no animal testing, engaging consumers through charity[9] and collaborative partnerships with groups dedicated to human rights and environmental conservation, the cosmetics retailer has established itself as a leader in the natural beauty market by staying true to its roots.

Consumers long for more companies that can help them better connect to family, community, and the environment; yet at the same time, brands find that consumers' actions are not always aligned with their words—that is, they are not doing their part, either. While 65 percent of consumers believe they can influence companies to do better by "voting" at the checkout counter, only 28 percent actually confirm having purchased products or services enhancing this new vision. Why?

One of the main reasons is cost. For example, if you compare the higher price of organic produce with the cost of conventional items, you will find that the price tag for organic can be cost-prohibitive. This is because the production and handling requirements for certified-organic items come at a higher cost to the producer, which comes through in market value. Similarly, so does upholding the standards for fair-trade factory conditions, making artisanal products by hand, and sourcing high-value, recyclable materials to make packaging.

Companies that care for people and the environment will often internalize costs currently externalized by their competitors. To make a profit on goods that cost more to make, producers must sell them at a higher price. This puts a price premium on the conscientious company's goods, which can put the consumer in a tough spot.

While consumers and end users still show demand for products that are convenient and inexpensive, they can buy only what is made available and accessible to them. Like with recycling solutions, if a product or service is cost-prohibitive, inconvenient, or difficult to understand, consumers are unlikely to use it. This spells opportunity for brands to evolve their engagement with customers and move beyond just selling more—to fostering active, loyal brand supporters in a circular economy.

The Dr. Bronner's brand of organic products, with its famous liquid soaps and peace-loving labels, has been around since 1948.[10] But longevity and brand loyalty are not won with time alone. All of Dr. Bronner's products are available at a price point comparable to or below other major organic brands in the category, many of which the company has sued for misleading labeling.[11] Accessibility is what makes the brand an easy choice, beloved for its concentrated, high-quality formulas (some applications need only a few drops, according to their cheat sheet)[12] and credo of sustainability.

Starting out in small health stores and since expanding to online sales and the shelves of mainstream groceries, Dr. Bronner's has remained relatively affordable while maintaining the progressive business practices and ethical supply chains that consumers have come to care about. In addition to its commitment to reusing and recycling in its facilities, the company uses 100 percent post-consumer recycled PET plastic bottles ("bottle-to-bottle") for all of its liquid and pump soaps; half the

Dr. Bronner's

11.4 Dr. Bronner's uses 100 percent post-consumer recycled plastic for all their liquid and pump soaps and commits to reusing and recycling in their facilities—business practices consumers care about.

bottles are made from locally sourced plastic resin collected from curbside pickups in California.[13] **SEE 11.4**

Relevant to packaging, the brand has an opinion on bio-plastics, expressing optimism for the future but concern that turning plants into plastic is still more energy intensive than recycling.[14] About this and other matters—such as regenerative agriculture, hemp policy reform, and vegan, organic, and fair-trade certification—Dr. Bronner's posts to its website, blog, and social media platforms. Providing access through information and a community of user-provided content, Dr. Bronner's shows consumers that it cares. In response to the brand's cultivation of transparency and culture, consumers care about the brand in turn.

Authentic Sustainability Is Nonnegotiable

Accessibility provided by the right price point is important, but sincerity is essential. Consumers are drawn to brands that keep them in mind and represent the things they care about. A brand that demonstrates consideration for its employees, suppliers, and community; the earth; and, intrinsically, the kind of company it wants to be earns the trust of its customers, who will be more likely to pay more for its products and follow its journey, wherever it may go.

Steve Jobs had it right when he famously said, "You can't just ask customers what they want and then give it to them. By the time you get it built, they'll want something new." Brands must be able to identify core insights about their customers— even before *they* do—and then innovate products, services, and programs to support evolving aspirations.

Many companies big and small have embraced **corporate social responsibility** initiatives to inspire brand affinity (and command premium prices) in the midst of an environmental zeitgeist. Highly discerning, consumers have a nose for **greenwashing** and are allergic to ethical claims that are inauthentic, vague, or misleading.[15] But in the digital era, consumers no longer have to take a brand's marketing or messaging at face value.

"There is a wealth of information out there that empowers people to know what they're consuming and how it affects their health and the environment," says Annie Jackson, vice president of merchandising and planning for Credo Beauty, a San Francisco–based retailer of clean beauty products. Carrying more than 100 clean beauty brands sourced from the West Coast all the way to New Zealand, Credo Beauty is dedicated to products that use ethically sourced ingredients and are cruelty-free and sustainably packaged, and it understands that sustainability in any industry today is nonnegotiable. "If you were to start formulating and creating a brand today," asks Jackson, "why would you do it any differently knowing the impact?"[16]

Gone are the days when being "environmentally friendly" automatically adds value in a largely niche market. More and more, consumers expect companies to dedicate themselves to making a positive social or environmental impact on society as a baseline, and they want to be able to trust them to prioritize ethics.[17] This trust is established through a brand's authentic stories, transparency, reputable certifications, and demonstration of measurable effects and by its doing most of the heavy lifting for consumers with regard to information and ease of use.

This opportunity is relevant to all industries. Today the relationships companies have with individuals are far more transactional and less about brand affinity than they could be. Every industry has room to improve. Unprompted, individuals identify companies such as Apple, Amazon, Google, Procter & Gamble, Trader Joe's, and Panera as leaders in the endeavor for a simpler and more connected life because they listen to people, generate keen insights, and invent new products, services, and experiences that create relationships and engage with the lives of their consumers.[18]

The business case for supporting the good life and a more circular economy is clear. Engaging with people on a deeper level increases customer loyalty, advocacy, and, ultimately, purchases. Failing to engage puts brands at an increasing disadvantage, while the opportunities presented by bringing better brands to consumers are almost limitless. Successful brands today understand the power of meaningful connections, and that intimacy, humanity, and trust—when delivered—supports abundance for both business and consumers.

CHAPTER 12

Designing for the New Consumer: Abundance without Waste

Raphael Bemporad
and
Liz Schroeter Courtney
BBMG

WHERE OWNERSHIP IS A SIGN OF SUCCESS, THE MORE one owns, the more important one is. We grow up to get a job so that we can buy a house, a car or two, fine clothes, and everything else we have been told will make us happy. Drawn in by advertising as soon as we are old enough to pester for the latest toy, the coolest cereal, we've experienced this perceived need to keep up with friends and peers. Like a goldfish expanding to fill a larger tank, the spaces we inhabit have their own way of convincing us to stock up and have *more*.

The pressure to own isn't purely human nature; there are architects at work—literally and figuratively. In the 1930s legendary public relations leader Edward Bernays convinced architects, contractors, and decorators to design built-in bookcases in new homes so that his clients, which included major book publishers like Simon & Schuster, could sell more books. This is a classic case of consumption driven by creating allure around the ownership of products that people didn't even know they needed, a model companies use to work consumers in the present day.

For much of our recent history, to borrow or rent has signified an inability to own; it came with a tinge of shame. The availability of credit in the 1980s ousted limitations and gave manufacturers little incentive to build products to last when consumers could now easily buy new products. A faster-moving world produced inexpensive packaging that made goods easier to buy and even easier to toss. **SEE 12.1**

This paradigm of ownership motivates us to consume with profound consequences and mounting challenges for our quality of life and the health of our planet. It's a vicious cycle that today saddles people with mortgages and credit card debt

Syda Productions/Shutterstock

12.1 The availability of credit makes goods easy to buy, while inexpensive packaging and mass production make them even easier to toss.

they cannot maintain (let alone afford) and piles of material goods that are rapidly depreciating or in need of constant maintenance.

The psychic weight of this world of ownership has many consumers yearning for a new model—or perhaps an old model that's ripe for rejuvenation. The *sharing economy* is not new. After all, communities have borrowed and shared for ages. But new technology and innovative approaches are making circular systems of access an appealing alternative to ownership.

A New Mind-Set: Abundance without Waste

The re-mainstreaming of sharing-economy concepts for *durable goods* signals high hopes for packaging systems that are better for the planet. Putting people at the center is the strength of today's most dynamic sharing apps. Applying this to packaging has the potential to unshackle consumers from current definitions of consumption, as well as producers from their current negative impacts.

Today a rising generation is experiencing ownership not as a benefit but as a burden that hinders their desire for freedom and meaningful relationships. Furthermore, the burden of ownership undermines desires for travel, more-fluid work, and nurturing relationships through sharing. **SEE 12.2**

Feeling the impact of terrorism, ongoing faraway wars, numerous economic crises, and environmental threats, an emerging twenty-first-century generation is experiencing an equal and opposite reaction toward generosity, creativity, collaboration, and caring. This generation of *aspirationals*

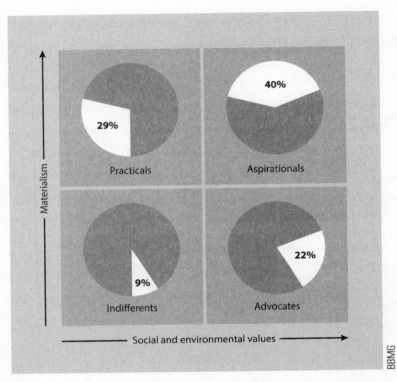

12.2 The aspirational consumer is redefining materialism along an axis of social and environmental values.

isn't defined by age but rather by the desire for their actions to meet their own needs, have a positive impact on others, and connect with an ideal or community bigger than themselves. Aspirationals are redefining the idea of abundance, seeking experiences that use fewer resources, produce less waste, and help them make deeper connections, leading the shift from a scarcity mentality to a new reality in which access equals prosperity.

When they do purchase material goods, aspirationals gravitate toward brands that reflect their values. For instance, a FEED Project tote bag signals concern for human rights and replaces an armful of disposable shopping bags. The brand S'well makes a popular water bottle that is functional and environmentally friendly, as reusing a durable bottle offsets the need for disposable plastics. Mason jars are making a comeback, used as breakfast smoothie containers and wedding reception decor by a generation longing for a simpler, less wasteful way.

Aspirational consumers may be rebelling against the ownership ideal that was championed by previous generations, but it's not only young people who are waking up to the negative consequences of yesterday's American Dream. In fact, global research[1] reveals that this aspirational mind-set spans multiple generations, and it is fueling the shift from an ownership economy to one driven by access, sharing, and collaboration.

In virtually every product category, including packaging, brands are meeting this aspiration with new business models and brand experiences that point the way to a new era of design, innovation, and impact.

A New Value Proposition: Peer-to-Peer Sharing

One of the ways that brands are rising to the occasion is through the flourishing of ***peer-to-peer businesses*** that facilitate human connections, authentic experiences, and flexible work opportunities. Community bulletin boards and local artisan fairs have always provided people with ways to rent out their extra room, hitch a ride, sell their handmade goods, or find a helping hand. In the digital age, savvy upstarts are fostering these connections with new levels of searchability, personalization, transparency, and trust.

New models for package delivery such as Nimber and Roadie do away with the need for bulky boxes and wasteful shipping materials, allowing users to send parcels along with drivers "going that way anyway." With app-based communities connecting people and drivers going in the right direction, the people can leverage existing resources to offset delivery routes and kill two birds with one stone. Available to small businesses and enterprises, these service models have implications for e-commerce and other packaged goods.

"People literally love to deal with people, because people trust people more than they trust companies," says Nimber CEO Ari Kestin.[2] Whereas packaging serves functions of quality control, product protection, and marketing in the consumer goods space, the human touch as offered by peer-to-peer and rental networks is a strength that can offset these functions—or even replace them entirely. **SEE 12.3**

The craft fair went digital in 2005 when Etsy launched an online marketplace for makers of handmade and vintage goods; as technology has helped the peer-to-peer model scale up, the humble crafting site has amassed more than 33.4 million active buyers as of 2017.[3]

Homeowners looking to make a little money on the side began renting to travelers seeking vacation homes on sites like VRBO and HomeAway in the early 2000s, while the especially thrifty could crash at a new acquaintance's pad via Couchsurfing. For people with skills to share, the marketplace

BBMG

12.3 Putting people at the center is the strength of today's most dynamic sharing apps.

for freelance work moved beyond the temp agency. There's TaskRabbit, an online marketplace where people can post their requests for help with everything from hanging a shelf to picking up dry cleaning to staffing a cocktail party; people with a knack for caring for furry friends can connect with peers on DogVacay.

Over time Airbnb took the home-sharing model to the next level by shining a light on the opportunity not just to save money but to have a more authentic travel experience—to enjoy a city like a local and bond with people who live there. The company has since expanded its offerings to include experiences—everything from wine tastings to surf lessons—curated and hosted by locals. Today Airbnb is a $30 billion business.[4]

Similarly, the peer-to-peer model has transformed the very idea of mobility and the transportation industry with the arrival of apps like Uber, Lyft, and Juno that combine the convenience of requesting a ride with your smartphone and the flexible side job of being a driver-for-hire on a schedule that you determine. And for those who want to share their car but not their driving skills, services like Getaround help connect idle cars to drivers in need of temporary wheels; a similar service called Spinlister allows people to lend out their bicycles and other sporting gear.

In the peer-to-peer marketplace, everyday people have become the new faces of brands. We are no longer merely buying products but forging human connections. The internet has made possible a vast array of networks for selling services, sharing space, and cutting out the middlemen who have been the economic gatekeepers for far too long.

Rental and Subscription: Access above All

The desire for access without the burden of ownership has also led to innovations in the *rental and subscription* realm. For decades we were content to leave rental to the world of movies and cars. Then in the early 2000s, there was an explosion of new subscription models that delivered access to a seemingly limitless inventory of shared items, from high-end fashion with Rent the Runway and Le Tote, to local personal transit with Zipcar and bike-sharing programs, to toy lending with a subscription to Pley.

Subscription services like Spotify and Netflix made content available on demand, while GPS-enabled mobile apps unlocked services like car2go, whereby rental cars are sprinkled conveniently across city streets rather than in an asphalt lot by the airport.

Packaging service RePack offers brands (including those on boutique and resale sites) an opportunity to offset e-commerce shipping waste with returnable and reusable

RePack

12.4 RePack is a simple, circular solution to the currently disposable packaging stream that makes online shopping and e-commerce possible.

shipping options for products they sell online. With incentives for consumers, it's a disruptive but simple solution that loops people and retailers together with a sustainable alternative to a ubiquitous waste stream. **SEE 12.4**

And in the growing gig economy, freelancers can still enjoy the communal and functional benefits of office life by joining a coworking space. The phenomenal growth of WeWork collaborative workspaces around the globe—248 locations in 58 cities at the time of this writing—shows how the rising freelance workforce, start-ups, and remote employees are adapting. Replicating office perks, coworking spaces have even become the preferred mode for many of today's freelance workers, who dig the topnotch coffee, speedy Wi-Fi, networking events, and comfy meeting spaces.

Today rental models aren't just about accessing goods when you cannot afford to own; it's the preference of consumers who desire more flexibility, more variety, and more abundance without the waste. Rewarding consumers with the right incentives takes that preference from an option to an essential.

Recommerce Models: Goods Go On

While access through peer-to-peer sharing and rental and subscription models has reduced our need to own, there are plenty of times when we still want to have and to hold something for keeps. To address the reality of limited material resources and to curb the amount of goods that end up in our landfills, there have been breakthrough innovations in *recommerce* that make owning a used product as good as, if not better than, buying it brand-new.

Thrifting has long been a popular pastime for savvy, budget-conscious shoppers who enjoy the thrill of the hunt. But sometimes you have a specific item in mind and no time to dig—not to mention the ever present fear of buying a lemon. In the mid-1990s, eBay and Craigslist helped connect buyers and sellers of secondhand goods. Sites like Tradesy and Kidizen have since established niche recommerce communities for goods like high-end fashion and children's items. **SEE 12.5**

One function of packaging is maintaining and communicating the "newness" or "freshness" of durable and fast-moving goods. The great value in recommerce is the ability to access and own gently used items in ways that are, quite literally, out of the box. There is less packaging than when buying new and, even better, when you're sharing with your neighbor and not buying something used off the internet, there's no packaging at all.

Helissa Grundemann/Shutterstock

12.5 Access to gently used secondhand goods through recommerce platforms reduces packaging waste by bringing products "out of the box."

Shining a light on the durability of their wares, Patagonia's Worn Wear program encourages customers to repair rather than replace their goods—or to trade them in for store credit, to be cleaned for someone else to enjoy. Patagonia and others are riding a growing wave of interest among consumers in buying clothes and household items that are built to last, inching away from the disposable lifestyle toward a new definition of convenience that empowers people to truly invest in their possessions with both time (repair and reuse) and money (your product).

The Takeaway: Three Opportunities for Brands

At a time when deeper connections, unique experiences, and expanded access are increasingly prized over merely accumulating more stuff, how can today's brands meet consumers where they are and continue to innovate for long-term relevance, resilience, and success?

First, the sharing economy enables the opportunity to save and maximize resources. Airbnb is a multibillion-dollar company and now the largest hotel brand in the world—and it has never had to own any property like its traditional competitors. Instead, Airbnb has tapped into existing real estate and unleashed the earning potential of thousands of homeowners across 65,000 cities in 191 countries.

Chip Conley, who served as head of global hospitality and strategy for Airbnb from 2013 to 2017, helped guide the brand's transition from renting rooms to delivering curated travel experiences. "When people are traveling, getting to know others and turning strangers into friends, we create a world where there are a lot fewer people who seem alien to us," says Conley.[5]

Second, sharing models provide companies with a deeper understanding of what consumers really value. Spotify's cofounder and CEO Daniel Ek came of age during the Napster craze, and he appreciated the thrill of online music discovery and unbridled access to an endless world of music. But illegal file sharing was leaving artists and record labels out of the value chain. "I wanted to do something where in the end the artists, the labels, and everyone in the ecosystem was a part of the development," says Ek.[6]

The brilliance of Spotify is how it re-creates traditional discovery channels (listening to the radio, trading mixtapes with friends, getting recommendations from a record shop clerk) with algorithms that serve up customized recommendations or click-of-a-button playlist sharing with friends—so users get all of the benefits without accumulating more physical media on a shelf. Spotify and other streaming platforms enable people to access a broader diversity of artists than ever before because there's no risk in trying something new when there's no waste and so much potential reward.

Third, these new models turn business barriers into innovative pathways for growth. Fashion brand Eileen Fisher's layers of flowy linen, earth tones, and knee-length cardigans have long been a favorite among middle-aged, upper-class women. But with an incredible commitment to circular design and ethical business practices, Eileen Fisher is breaking through to a younger, aspirational audience for whom sustainability is a major appeal but who once found the brand's initial price point out of reach.

To meet its goal of recycling 1 million garments by 2020, the brand launched Renew, a program that gives Eileen Fisher clothes new life. Customers bring back old pieces, and Renew finds them a new home or turns them into entirely new designs that are sold online and at select Eileen Fisher retail stores— at lower prices than new garments. Since launching Renew in 2009, Eileen Fisher has taken back 800,000 garments, for $10 million in resale revenue, including $2 million donated to charity.[7]

Seeing barriers as opportunities for growth continues to inspire the brand's design processes for new products. Cynthia

Power, who helps lead the Renew program, says the company envisions a future in which waste is a thing of the past, and circular thinking makes it all possible.[8]

Fueled by an aspirational mind-set and the rapid evolution of new technologies, today's consumers want access to better experiences with less waste. Reimagining consumption models yields high-value products and services that can generate new sales, improve customer loyalty, and integrate simple sustainability solutions for profitable progress overall.

Where today's packaging of goods has enabled a faster-moving world that automates shipping, handling, and use, the sharing economy brings *people* back to the center. What functions does packaging truly serve, and how can we satisfy these goals in ways that are friendlier for people and less costly to the environment? The answers to these questions will inform the designs for the new consumer.

Brands that leverage new business models and innovation platforms will win in a world of limited material resources and limitless creativity, helping move us as a society toward more-circular systems of consumption, resilience for business, and a positive impact for people and the planet we share.

Changing the Paradigm to Enable and Inspire Responsible Consumption

Virginie Helias

Vice President, Global Sustainability, Procter & Gamble

WHILE NEW SHARING AND SERVICE MODELS OFFER access with less waste, we are also beginning to see this kind of change within the CPG space. We know that lighter, cheaper, and more portable and affordable material alternatives have allowed manufacturers to break down barriers of cost for consumers. Having increased consumers' access to affordable goods, companies now have the responsibility to innovate for the long term, even as the pressure to deliver profits—*to grow*—only mounts. **SEE 13.1**

To grow, brands often ask themselves, *How do we drive more consumption?* The answer has been simple: by defining for consumers what a good life should look like and by offering the products they should aspire to in order to successfully live it. It's a cycle that has locked companies big and small into continually striving to sell more—sometimes at all costs.

Fast-moving goods aside, even some "durable" goods are designed to break or become obsolete—the policy of

13.1 This Tide advertisement from 1948 is an example of a brand's long history of driving consumption.

planned obsolescence. Electronic technologies upgrade every few months. Accessories and operating systems slow down and become difficult to manage without updates. Consumers have become so accustomed to and even dependent on these technologies that, however begrudgingly, they intuitively buy upgrades to retain their functionality.

Under pressure to deliver the most total shareholder return, companies generate value through the quantity of goods and services they sell. Producers have designed a vision for a good life that today nurtures a disposable society—making it all about a life defined by a journey of purchases that can lose their meaning and quite quickly be thrown away.

This linear, one-way pattern is increasingly wasteful, leaves consumers unfulfilled, and holds industry back.

Through the promotion and enabling of a different version of a good life, we can reevaluate consumption to change the paradigm around profitability and business growth. Brands that give back to society are the better choice and will thrive, while those that do not give back risk irrelevance and extinction.

Connecting to Better Brands

It is the role of industry to be the bridge between sustainability and the business units that reach consumers. For us as producers and designers in the current culture of consumption, educating ourselves is the first order. For example, when a recent award-winning environmental documentary came out in theaters, my team and I at Procter & Gamble went to the theater together. I was very touched by it, and I sought to understand the science behind sustainability. I learned that, from a life-cycle assessment standpoint for detergent, 80 percent of the carbon footprint comes from the power needed to heat the water in your washing machine; I would never have guessed.

Education, however, is not the primary way to convince consumers about sustainability: the key is to make sustainability desirable so that people embrace it because they *want* to, not because they have to or feel guilted into it. Our two key levers for that are innovation and aspiration. An example of *innovation* is P&G's gel laundry detergent, which works in cold water (it starts working at 15 degrees centigrade) and it is compacted, saving both energy and packaging resources.

Aspiration is about making sustainability what people want to do. If you have to convince people to do something, you have failed on irresistibility. You need to help them fall in love with it and seek it out.

At a recent conference, we asked consumers, "What makes you happy? What do you truly value?" The answers were all related to staying connected:

- **Connected to oneself**—making time to nurture oneself physically, emotionally, intellectually, and spiritually

- **Connected to others**—making time for family and friends to create memories

- **Connected to the world**—making time to be con-nected with nature and communities[1]

Products and packaging that require less time, energy, and resources to consume bring people closer to the values they hold dear. Making these choices irresistible to consumers drives value for business through meaningful connection and redefines what it means to grow as an industry.

As brands it is our role to promote a lifestyle that is bet-ter for people, better for the world, and much closer to these consumer aspirations. That is why we must ask a new question:

not just *How can we drive consumption?* but *How can we drive* **responsible** *consumption?*

These examples from the sustainability journey of P&G, one of the largest consumer products companies in the world, are not about *more* consumption but rather *better* consumption. Through innovations and communications, we strive for a paradigm with no trade-off between delivering shareholder value and connecting consumers to the things they value.

Turning Laundry Time into Turbo Time

Whether it's water, energy, or packaging, wasting resources is becoming less and less acceptable to people. While *saving money* and *health and well-being* are top considerations when purchasing a product, consumers also cite as important considerations products that help them produce less waste and that limit their consumption of water and energy.

At P&G we designed Tide Turbo laundry detergent for high-efficiency machines with quickly collapsing "smart suds" that not only increase the friction necessary for effective stain removal but rinse clean more rapidly. Reducing the need for long rinse cycles, the formula saves up to 25 minutes of wash time for up to 10 full gallons of water per wash versus other detergents. That's 25 fewer minutes of powering the washer—shortening each cycle for less energy and materials used and potentially extending the life of the machine.

Reputable certifications are a significant way to connect with consumers and establish trust in the authenticity of your product. Tide's Purclean line is formulated with 65 percent bioderived materials and is the first plant-based liquid laundry detergent certified by the US Department of Agriculture. To attain the USDA BioPreferred seal, the product must be at least 34 percent plant based, so the current formulation contains nearly twice the plant-based content required for certification.

Produced in **ZMWTL** facilities powered by 100 percent renewable wind electricity and steam power, the plant-based Purclean formulation can be used in cold water with the same performance as Tide. It offers an option for consumers seeking a simple step to integrate renewable plant-based products into their daily routines that meet their cleaning expectations.

A growing number of consumers are seeking out natural ingredients in their products. Thus we are exploring increased use of renewable materials. Providing consumers with better versions of the products they use every day drives not more consumption but better consumption. This links consumers to brands on a positive loyalty loop that keeps them coming back. After all, more-sustainable products can only have an impact if people use them!

Can Anything of Value Really Come out of a Diaper?

Parents, siblings, friends, and babysitters know: babies' diapers are changed five or more times per day. In both industrialized and developing countries, disposable diapers have almost completely replaced the reusable cloth versions of yore. An average child will go through hundreds of diapers in their lifetime. Today without a recycling solution, it's a waste stream that puts a burden on landfill sites, releasing greenhouse gases that contribute to climate change.

Can anything of value really come out of a used diaper? Our Pampers brand has committed to recycling the unrecyclable. Let that "stink" in for a moment. Developed and patented by Fater S.p.A., a joint venture set up by P&G, breakthrough technology is currently generating high-value secondary materials, such as plastic, cellulose, and absorbing material, from used diapers as we work to advance it so that it can be deployed at scale.[2] The world's first industrial-scale plant capable of recycling virtually 100 percent of used absorbent hygiene products is currently operating in Italy, with a second one in the Netherlands to follow.

In developing countries where waste infrastructure is limited and open landfills and littering pose a major societal and health challenge, diaper recycling will be a huge positive. One metric ton of **AHP** (or *absorbent hygiene products*—the category name for baby diapers, sanitary protection pads, tampons, adult incontinence products, and personal care wipes) waste, after being separated from the organics, yields around 150 kilograms of cellulose, 75 kilograms of mixed plastic, and 75 kilograms of absorbing material. These can be used in new products and processes such as biofuels, school desks, and gardening equipment.

Partnering with millions of parents around the world to reduce the impact and improve the life cycle of an essential product, the diaper-recycling initiative won the European Commission's Circular Economy Champion prize for Fater,[3] as well as an EMBRACED grant (which stands for *Establishing a Multipurpose Biorefinery for the Recycling of the organic content of AHP waste in Circular Economy Domain*).

Consumers implicitly know that a recycling solution is an improvement over the current way diapers are consumed. Researching, developing, investing in, and supporting solutions for common waste streams associated with your product is a statement of design that consumers will recognize and your industry will emulate.

Cleaner Hair, Cleaner Oceans

If designed outside the capability of resource management systems, plastic packaging makes its way into the environment and even our food chain. As intimidating as some sustainability and environmental problems can be, this pollution is an issue that is very easy for consumers to understand. Consumers look to brands to provide ways to reduce the amount of waste littering land and sea.

We thought the world's number one shampoo brand, Head & Shoulders, had the potential to reach the most people on the issue of ocean and marine plastic. The product is already intended to clean and care for the hair of people around the world—could it also clean and care for the ocean? **SEE 13.2**

We created the world's first fully recyclable shampoo bottle made with beach plastic for Head & Shoulders[4] to highlight the issue of ocean plastic. The bottle was produced and made possible by collaboration with recycling company Terra-Cycle and waste management firm SUEZ. Sourced through partnerships with beach cleanup organizations already picking up litter on the shores of oceans and other waterways, ocean plastic originally headed for landfills was used to establish a new supply chain.

Procter & Gamble

13.2 The world's first fully recyclable shampoo bottle made with beach plastic established a new supply chain and highlighted the issue of plastic pollution.

P&G's retailer partner Carrefour helped engage shoppers at the point of sale, raising awareness about plastic pollution in the ocean and the role that people can play by recycling. Consumers not only connected with the concept of a package made of a waste stream they are familiar with but were encouraged to recycle by the story of what it took to produce the bottle. The change from the signature white bottle to gray is a visual marker for the integration of beach plastic, which comprises everything from water bottles and toys to food packaging and any other plastic that washes up onshore.

The breakthrough project was the first of its kind— the largest production run of recyclable bottles made with post-consumer recycled beach plastic—and would go on to win a Momentum for Change Lighthouse Activity Award from

Procter & Gamble

13.3 Launched at the World Economic Forum in 2017, the
Head & Shoulders beach plastic bottle would go on to win a
Momentum for Change Lighthouse Activity Award from the
United Nations.

the United Nations.[5] Inspired by the response and humbled to
lead the conversation, P&G has since moved forward to integrate beach plastic across several product lines and countries,
using even more of the plastic that had littered the world's
shores. **SEE 13.3**

Sharing the Load

Back when the "Cool Clean" campaign for Ariel, our global
laundry care line, was launched, the team knew we had a
unique technology that allowed people to effectively wash their
clothes in cold water—a big plus for energy conservation. We
leveraged this in our brand communication, homing in on the
fact that if enough people partnered with us and played their

Procter & Gamble

13.4 Ariel's "Share the Load" campaign went viral in 22 countries and 16 languages and won a Glass Lion at the Cannes Lions International Festival of Creativity.

part, the power saved could light up 1,000 villages in France for one year.

Brands can operate in the interest of more than one social problem at the same time, connecting with consumers on multiple levels for maximum impact. Research shows that 76 percent of Indian men feel that laundry is a woman's job, and 85 percent of Indian women feel that they are working two jobs—one at work and another at home.[6] "Share the Load" is another Ariel message that promotes gender equality in households. **SEE 13.4**

With an international campaign that started in India, where 95 percent of households surveyed view laundry as exclusively a woman's job, Ariel started a conversation that went viral in 22 countries and 16 languages[7] and won a Glass Lion at the Cannes Lions International Festival of Creativity. A host of celebrities and other famous faces have showed their support for the campaign, including a top executive from one of the world's largest social media platforms,[8] showing that a more equal world would be a better world for all of us.

Addressing the root cause as the cycle of prejudice passed down from one generation to the next, this award-winning campaign got 2 million men to pledge to #SharetheLoad. That's 2 million more people invested in the brand—and its mission to change the paradigm around both gender equality and resource management.

There is one thing stronger than all the armies in the world, and that is an idea whose time has come.

—Victor Hugo

A *good life* defined by health, abundance, and freedom, with less negative consumption and waste, is one of those ideas. Brand innovation is necessary to *design* this good life, but it is not sufficient to get consumers to bite. Companies are responsible for *delivering* this innovation through brands that make responsible consumption possible and irresistible.

Brands can drive the conversations that help people better connect to their real selves. Making this case can empower companies to overcome the prohibitive barriers that get in the way of invention and forward thinking. Businesses benefit from listening to consumers and working to reach them, touching them with products that enhance their everyday lives. Those that do will be rewarded in the long run, making an impact that goes far beyond business.

CHAPTER 14

Value for Business in the Circular Economy

Lisa Jennings

Vice President, Global Hair
Acceleration, Procter & Gamble

A CHANGE IN SUPPLY CHAIN PRESENTS A CHALLENGE FOR any business. This is particularly true when changing from a linear system to a circular one—where we aim to recover and reuse materials previously considered waste. In a circular system for recycling plastics, no single group is entirely responsible. Our collective challenge is to ensure that the recycling infrastructure includes access to collection systems, enables participation by consumers, and supports end markets for recycled materials.

The linear production economy has long been effective in meeting consumers' needs, creating jobs, delivering profits, and generating growth. As many companies strive toward a more circular economy, with aims to build business while providing environmental benefits, however, we see shortcomings. Even the best programs for recycling and waste disposal can be improved, and most people (save very few conscious consumers) will not voluntarily pay auxiliary fees to supplement these systems.

With P&G's Head & Shoulders beach plastic initiative, we chose to confront these challenges and show how we care for consumers and the environment. We aimed for and delivered a circular supply chain while building our business. We faced challenges, but with every step away from business as usual we made one in the right direction toward new solutions in a more regenerative, restorative system. We innovated over points of resistance to allow for a more circular economy and to champion and effect change within our organization and beyond.

Small Steps Are Still Steps

Developing products that consumers want to buy is any company's goal—and packaging makes it possible. But for business to scale for the long term, it must do so in a way that considers resources and works to change the current unsustainable system. Advancing a circular economy means creating a market for recycled material by infusing packaging streams with recycled goods and designing for reuse and recyclability. Plain and simple.

Understanding this is essential to starting conversations around the circular economy. Do your research on policy frameworks, emerging models, and the priorities in your company's culture. Talking about potentially game-changing initiatives is half the battle. From casual discussions with colleagues to the formal proposals presented to superiors, sustainability knowledge makes you an asset, and your belief in its implications can help convert naysayers and embolden fellow stewards in the community.

The reality is that businesses can and should play a critical role in driving and inspiring change in the world. This is what inspired me to make sustainability a priority for my business unit: Procter & Gamble hair care. In the beginning there were only aspirations for what could be the largest and potentially the most meaningful solution for discarded beach plastic in the world to date—and a first for the hair care category overall. The rest was a feat of collaboration, dedication, and innovation that catalyzed a shared vision for a change in the way we produce, consume, and dispose of packaging. **SEE 14.1**

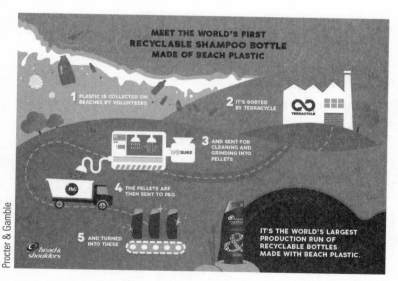

14.1 The world's largest (and first) production run of recyclable bottles made from beach plastic started with a conversation. The rest was a feat of dedication.

Having worked for P&G for more than two decades, starting out as an assistant brand manager for the cosmetics and skin care units and landing in lead marketing roles, I knew the obstacles we would face both internally and externally. A conversation, backed up by research and diligence, was the first step in any initiative pursued.

A *circular economy* is one that focuses on durability, recyclability, and use of renewable resources, including energy *inputs*. Shining a light on practical and scalable efforts for sustainability stands out in an industry that still profits from the status quo. As brand owners, we are in a good position to inspire consumers and work across the supply chain to spur a shift that favors the use of more-sustainable packaging material and supports the improvement of recycling systems for efficient material sourcing.

For example, the European Union's plastic strategy calls for binding targets for collection, sorting, and recycling,[1] but it also calls on industry to strengthen the economics and quality of plastics recycling by improving product design, boosting recycled content, and refining the collection and separation of plastic waste. This call to action is, in fact, incentivized by policy frameworks to drive investments and innovation for the plastics value chain.

Familiarizing yourself with the resources at your disposal is a powerful step in developing actionable proposals. Contacting public offices is one way to do this. Leveraging your knowledge of the policies and framework resources improves your ability to steward change in your organization. As you get started on your circular-economy journey, small steps are still steps, and these largely consist of conversations that build momentum as you inspire people to pay this information forward.

Sustainability Stands Out from the Status Quo

There are good examples of effective *voluntary producer responsibility* (**VPR**) programs that businesses and consumers can emulate and customize to work for them. Let's switch from plastic for a moment and consider aluminum recycling. Aluminum can be reused and is, to date, one of the most recycled materials in the world. In fact, aluminum recycling played a role in supporting the expansion of recycling to include plastics.[2] Yet most aluminum beverage capsules are designed for single-use and considered disposable, a ubiquitous example of typically unrecyclable items made of a highly recyclable material.

As part of its capsule recovery program, Nespresso's "Positive Cup" initiative commits to expanding the company's capacity to collect used capsules in 100 percent of the places where it does business, creating partnerships with UPS in the United States and at thousands of corner shops in the United Kingdom; it collected directly from the postbox in Switzerland and 15 other countries.[3]

This VPR supply-chain innovation not only provides local industries with access to recycled aluminum but also creates a virtuous end-to-end loyalty loop, locking consumers into the brand with convenient home delivery and the positive choice to recycle; this gives customers incentive to resist alternatives available in supermarkets—most of which are made from mixed materials and are therefore much harder to recycle. Keurig is another coffee manufacturer that has implemented VPR recovery programs.[4] It recently moved its Canadian business away from multilayer to recyclable polypropylene pods. Both companies are making efforts to address complex recovery situations.

Developing close collaborations across the supply chain puts you in a strong position to discuss how you and your suppliers can work together to offer a higher-value product to consumers. These dialogues add momentum to internal discussions and yield more-complete information for your colleagues to consider. Insights from local authorities and governments also help with this. Working out an arrangement designed to progress with returns or cost savings starts your proposals with big positives.

There are also ways to work recycling into your products and packaging at low risk with a highly sustainable return on investment. P&G's international Febreze brand works with TerraCycle to offer customers around the world free programs for municipally unrecyclable packaging. By paying Terra-Cycle to manage collection, processing, and logistics, Febreze employs VPR for its packaging as it continues working to support the public recycling infrastructure. Coincidentally, TerraCycle also manages Nespresso's recycling programs in Canada, Australia, and New Zealand. **SEE 14.2**

We have a way to go to achieve the ideal packaging management system. The ideal—the lighthouse vision—is 100 percent access to recovery systems for discarded packaging,

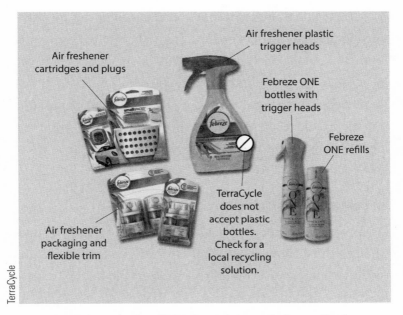

Air freshener plastic trigger heads

Air freshener cartridges and plugs

Febreze ONE bottles with trigger heads

Febreze ONE refills

TerraCycle does not accept plastic bottles. Check for a local recycling solution.

Air freshener packaging and flexible trim

TerraCycle

14.2 P&G's Febreze brand works with TerraCycle to offer free programs for municipally unreclyclable packaging to customers around the world.

100 percent participation in systems by consumers, 100 percent separation of materials, and 100 percent end markets for the recovered materials. Keeping materials in the supply chain, at high utility in the closed system that is the earth, is the work.

Consumers will vote with their wallets to support the businesses that take a longer-term view on resources only when businesses themselves believe that this is the right way to go. They must be inspired to be part of the generation of aspirational companies that devise a *new* business as usual. It is this belief that drives conversations and makes crystal clear the best next move on the journey to sustainability.

The value of better resource management can be redefined to mean more than dollar signs in the circular economy. From the largest global corporations to the smallest Main Street businesses, the most important assets are consumers. Without them we are nothing. By working to design our packaging into the circular economy, we can be the brands that mean the most to people—50 years from now and two centuries from now—for the work we do today.

An Ocean of Opportunities

Ocean and marine pollution doesn't necessarily come to mind when we think about poverty, but the truth is, all infrastructure issues are a matter of economics, and so are most social issues. Resource mismanagement ties back to *resource insecurity*— or not having enough to go around. By working to solve our issues with waste, we work to eliminate poverty and ensure the sustainable development of communities around the world.

As part of its aforementioned circular-economy package, the European Commission in 2015 presented an action plan in which it pledged specifically to take action to reduce marine litter with a view to implement the 2030 Sustainable Development Goals, 17 of which were created by the United Nations with the unifying thread to eradicate poverty. Better infrastructure, more access to public programs, and a more level playing field for all citizens are aspects of this.

With this top-down incentive, P&G's award-winning beach plastic shampoo bottle came about as a brainstorm to address this pledge. For me it was kismet. I was always personally very passionate about recycling both at home and at work.

Nicolas Guerrier/Procter & Gamble

14.3 Learning about the scale and impact of the issue of plastic pollution spurred the call to prioritize sustainability at Procter & Gamble.

Around this time the Race for Water Foundation—an NGO dedicated to identifying, promoting, and helping implement plastic waste recovery solutions to prevent their entry into waterways—showed me the scale and impact of the issue of plastic pollution. **SEE 14.3**

Conversations at the World Economic Forum annual meeting that same year made me realize that the problem of plastic pollution cannot be solved in silos. We all need to be part of the solution, leading by example, as both citizens and business leaders. To make P&G's sustainability vision a reality, we established a whole ecosystem of external partnerships that bring together committed and determined individuals and businesses.

In our passion to drive responsible consumption among our consumers, my team and I were able to have one-on-one conversations with different members of the company, both on the junior rungs and in leadership. The purpose was to get everyone involved to establish a long-term vision, set short-term goals, and ensure that we stayed on track. Full support from leadership was integral to get the project off the ground, but every department in the company was looped in to most effectively push through measures for maximum impact.

If this project were to be a success, sharing the model to inspire the business world at large would be the next step. While large companies have the resources and funding to take on a lighthouse project like this, smaller businesses have the flexibility to design for sustainability in the now. Embrace this freedom in conversations with your local governments, suppliers, and potential collaborators to come up with solutions for your company that turn out the most value for you and the most aspects of the proper packaging management system you are working to improve.

With regard to making the Head & Shoulders beach plastic bottle, we faced three key challenges. The first was availability of beach plastic that we could use in our products. Even though ocean and beach plastic pollution is a huge problem affecting the whole world, the quantities of beach plastic that were not degraded beyond the point of repurpose were limited. This is where our partners—TerraCycle and SUEZ, with the help of thousands of volunteers—played a key role in collecting, sorting, and ultimately delivering high-quality pellets made from available beach plastic waste. **SEE 14.4**

Nicolas Guerrier/Procter & Gamble

14.4 Beach cleanup organizations and NGOs were integral to building up a new supply chain for the Head & Shoulders project.

Sourcing relied on unconventional supply chains: hand-collection by a myriad of dedicated NGOs. TerraCycle, which specializes in the coordination of programs that collect and source difficult-to-recycle packaging materials for use in new production, was the perfect fit to mobilize consumers and make them part of the supply-chain process.

The NGOs and other organizations that came on board with us were happy to see the plastic given a second life rather than being shifted from the coast to the landfill, and we did not take this for granted. The beach cleanup organizations were already picking up litter from shores and waterways, but we were asking them to collect the beach plastic and, rather than place that typically unrecyclable material in the garbage, send it to us. This was a significant amount of work, and the physical beach cleanup efforts were completely voluntary.

Taking the time to explain what we were doing with the collected beach plastic, and why it was important, mobilized

Nicolas Guerrier/Procter & Gamble

14.5 Benoit Schumann, founder of Project Rescue Ocean, and Lisa Jennings of P&G.

Nicolas Guerrier/Procter & Gamble

14.6 Project Rescue Ocean is an NGO that drives awareness about the issue of plastic pollution; it is one of P&G's partners in sourcing beach plastic for the Head & Shoulders bottle.

the labor necessary to establish this new supply chain. Once people understood that they could unlock tremendous opportunities in reverse logistics, our NGO partners came through with enthusiasm and tireless dedication. Ocean and shore plastic is an environmental problem they connected to, so they helped P&G integrate it into our packaging—and they continue to supply our production to this day. **SEE 14.5 AND 14.6**

The second challenge was technical, as working with beach plastic is unfamiliar. TerraCycle collaborated with SUEZ, perhaps the world's largest sales force dedicated to the circular economy, to haul, sort, and clean the recovered beach plastic to supply resin for the production of the bottles. This aspect required sorting and cleaning off sand, organics, and residues multiple times. The compounding of beach plastics to ensure a homogeneous mixture of resins with the requisite technical properties was another step. **SEE 14.7**

Designing in this way is not the path of least resistance. To make a bottle with beach plastic, you need to ensure that the packaging is strong enough to prevent it from breaking. We needed to ensure that the shampoo inside the bottle was not compromised. This is where our technical teams and our packaging partner, ALPLA—a world leader in the development and production of plastic packaging solutions—played a key role. The bottle has three layers: the first inner layer is virgin resin, to ensure the right environment for the shampoo; the second inner layer is beach plastic; and the outer layer is virgin resin.

Innovative blow-molding techniques for the multilayer bottle needed for this project tested a novel process that has endured for new applications. Recycled beach plastic collected

Sandra Blaser/World Economic Forum

14.7 TerraCycle's Tom Szaky holds a mixed bag of plastic items found on the beach, including sand toys, goggles, and packaging—all different resin types damaged by sun and sand.

on the shores of oceans and other waterways is subject to UV and salt damage, making it more brittle than conventional rPET or rHDPE. In the design of the new bottle, performance tests (such as *drop testing*—simulating a drop from the top shelf of a storage facility) led to a last-minute bottleneck redesign by a hugely committed engineering team to ensure that the product could pass our high internal standards.

The third challenge we faced was one of marketing and branding. The Head & Shoulders bottle is recognizable because

14.8 Beach plastic, when mixed, becomes gray, so the Head & Shoulders bottle looks very different than the signature white— but it draws attention to its origins as "waste," driving value for consumers.

Procter & Gamble

of its iconic white-and-blue packaging. Beach plastic comes in a variety of colors, and when mixed it becomes gray; therefore the beach plastic bottle looks very different from the signature white and blue. The decision to go with a gray bottle was not an easy one, but we felt that it was the right thing to do. It allowed us to acknowledge the origin of the packaging material and talk about its authenticity. **SEE 14.8**

Establishing a new supply chain of previously unrecyclable material, as well as a product that could reenter the value system, was a process of moving parts. The first run of the beach plastic bottle required extra production time versus regular plant speed, as fragile material behaves differently on high-speed lines. Beach plastic is currently more expensive than virgin, as it is hand-collected by volunteers (it's practically artisanal) and because building the new supply chain took time.

We saw these investments yield real returns. The announcement of the Head & Shoulders launch at the World Economic Forum in Davos had hundreds of media placements (more than 4 billion impressions), and the push for our 100 percent PCR Fairy bottle less than a year later kept this momentum going. Our customers—the stores and retailers that carry the Head & Shoulders and Fairy products—work with us to support and promote the efforts through shopper

Nicolas Guerrier/Procter & Gamble

14.9 Engaging stores, retailers, and end users around the beach plastic problem drives value for P&G, which is cited by consumers as part of the solution.

marketing. Consumers are buying the packages, and conversations about the beach plastic problem cite P&G as part of the solution. **SEE 14.9**

The Head & Shoulders beach plastic bottle project was part one of a two-part message: The first part is that all plastic—even beach plastic—has value; using beach plastic in our packaging is a way to signal this value. The second part is the longer-term message of recycling. P&G hair care has committed to use 25 percent PCR plastic across the majority (more than 90 percent) of its entire European bottle lineup by the end of 2018. Consumer participation in recycling is needed to supply this PCR material, so we ask you: *please recycle!*

Casting a Wide Net

An ideal packaging management system consists of access to collection systems, participation by consumers, proper separation, and support of end markets for secondary materials. These aspects work in direct correlation with each region's respective biases about cost-efficiency, investment, legislation, and environmental concerns—areas affected by constant change.

Along with government involvement and investments in infrastructure, improvements to the current system require continued innovation in technology and the nurturing of ideas. Companies must make the bold decision to carve out their place in the future of sustainability and create the value they wish to see in the world. The public sector today provides incentives for designing into the circular economy. The private sector can continue to improve these systems to reward more companies for that action.

Leading this type of change is a challenge. New inventions, new manufacturing, and new supply chains are all hard work. But the payoff is simple: we get to stay in business. Efficient shepherding of the earth's finite resources is a business imperative. If we can't stop the world's current trajectory of climate change, what businesses will be left?

Industry is tasked with balancing the need for expected growth at a fast-moving consumer goods company with environmental responsibility. It is not just consumers and end users with problems to solve; manufacturers and brands also look for support within the shifting sustainability paradigm.

As consumers themselves, businesses look to their suppliers, vendors, governments, peers, and stakeholders for better materials, systems, guidance, and replicable models for designing for circularity in the face of many challenges. Thus the most important shift toward circular design is collaboration with valued experts. In fact, one of the most valuable aspects to come out of these initiatives is new partnerships.

Too often businesses operate as islands and miss out on opportunities to learn and grow. Whereas the externalities of a consumption-driven world are currently dispersed, a collective effort to better protect the earth's finite resources shares the weight of responsibility—sustainably. We hope that closing the loop in the business system by sharing models and innovating for new ones will inspire new conversations and more businesses wishing to start their circular-economy journey.

CHAPTER 15

The Future of Packaging

Tom Szaky

Founder and CEO, TerraCycle

UNDERSTANDING HOW WE ARRIVED AT A WORLD WHERE we as individuals are first viewed as consumers (before citizens) and actively fuel our economy through the consumption of larger and larger volumes of cheaper and cheaper products is essential to retracing our steps back in history and toward a circular economy. Since the debut of our disposable, throwaway culture in the 1950s, every lifestyle innovation has been centered on how to make products less expensive and more convenient. Throughout this process all stakeholders—from the producers to ourselves as consumers—have avoided paying the price for the externalities this linear, one-way economy brings, passing the entire brunt of that expense to our planet and future generations.

Today we have a very important choice to make: We can continue on the current trajectory and watch our planet and its environment push back with extreme weather events of increasing frequency and ferocity, which create climate refugees whose movements polarize our politics toward instability. We can go on voluntarily pumping the rampant pollution into our air, water, and food and continue to reduce our life expectancies.

Or we can start on a different route. We can move away from our current linear systems, bend them into circles of recycling and reuse, and work to make those circles as tight as possible by reducing the energy and resources needed to get our packaging from useless to useful again. To do this we must look to the wisdom all around us, both in nature and in our past.

To lay the groundwork for the future of packaging, we must take our first steps today. These first steps should be to make products and packages that are easily and locally recyclable. It is always easier and most efficient to design into today's recycling systems—no matter how flawed they may be—versus hoping that the infrastructure will rise to design itself into your innovations.

This journey should start with a conversation with your local recycler. Ask about the material they want to see in the waste they accept versus what they can technically process. You will likely hear that they would like to see valuable materials that are clear versus colored and opaque, white versus dark, rigid versus flexible, and not attached to (or composed with) other materials—simply, those materials that are currently easy to process into high-value outputs.

As you embark on this journey, dream about the future. Divorce yourself from your deepest assumptions. What if profit maximization weren't the central objective and cost weren't an issue? In such a reality, what types of packages or products would you design? And as consumers, what would you buy? Would they still be cheaply made and single-use, or would they be made from amazing, high-quality materials and have impressive functions and aesthetics?

If not recyclable, would products be durable, and could they be reusable? Perhaps packages for fast-moving goods would no longer have to be owned by the consumer and instead perpetually be the property and responsibility of the companies that made and sold them. Maybe the design of products and packages would go beyond the physical construction of an item to encompass the delivery and reclamation systems that maintain the flow of goods. And if packaging moved from being a throwaway cost to an asset of the manufacturer, couldn't all this end up being less expensive as well?

The role of business—of the major organizations, retailers, and consumer products companies—is to get for us, the consumers, what we want in the best and most affordable way. The products are a reflection of our desires. So when we as consumers desire things that generate less waste or are more circular, companies will produce more of those things.

The most important and empowering takeaway is that individuals have the power to quickly manifest this reality by voting for the future we want to live in—not just the one vote we cast every so often for our political leaders but the multiple votes we cast every day as consumers. As consumers we don't give ourselves enough credit for how powerful we really are.

The next time you go shopping, think of this awesome power in the following way: of everything you buy, tomorrow two more will be there; and of everything you don't buy, one fewer will be there. View your purchases as having a direct impact on the goods and services companies choose to make. If you want to eliminate waste in your life—and in the world— the answers will always come down to one simple thing: *consume differently.*

Glossary

Glossary terms are identified in the text in **_bold italic_**.

access One of four factors that affect the recycling of an item in the current infrastructure; it is necessary for consumer participation in recycling.

AHP Stands for _absorbent hygiene products_—the category name for baby diapers, sanitary protection pads, tampons, adult incontinence products, and personal care wipes.

aseptic carton An evolution from the traditional paper carton, the shelf-stable, multimaterial package extends shelf life for many food and beverage items; a well-known example is the juice box. Favored by manufacturers for their light weight and versatility, some aseptic cartons can keep products sterile up to one year without refrigeration. Several layers of polyethylene (PE, or #1 plastic), paper, and aluminum foils render these difficult to recycle, as the materials require separating, but some municipalities today accept them. _See also_ multi-compositional.

aspirational recycling The act of placing waste items with recyclables because one wishes they were accepted for recycling when they may not be. Sources of contamination for recycling bales—common items such as greasy pizza boxes, coffee cups, and plastic bags—cause problems for waste managers, affect the purity of secondary-material batches, and prevent bales from being sold for new production. Also known as _wish-cycling_.

aspirationals A new generation defined by the desire for their actions to meet their own needs, have a positive impact on others, and connect with an ideal or community bigger than themselves.

Aspirationals are fueling the shift from an ownership economy to one driven by access, sharing, and collaboration and from a scarcity mentality to a new reality in which access equals prosperity.

augmented reality (AR) Technology used to complement a person's environment through the addition of digital content or objects into their field of view, often enabled by a smartphone or tablet application.

Bakelite The world's first synthetic, durable plastic, developed in 1907 by Leo H. Baekeland. Long favored for high-value, expensive items such as wartime equipment and home appliances, it was adopted in the mid–twentieth century to the mass production of plastics.

balanced simplicity Living a simpler, healthier life; a significant factor in defining a good life for today's consumer.

biodegradable Describes a material that can be decomposed by living microbes. Generally referring to organic material that under the right conditions will completely biologically break down into carbon dioxide, water, and compost—elements that can be readily digested by the surrounding environment—the term is also used, perhaps misleadingly, to describe a type of bioplastic.

biodegradable bioplastics Bioplastics that, when discarded, are intended to biologically break down. In most cases biodegradable bioplastics break down only under the right conditions: a controlled, high-temperature industrial composting facility—not when littered, thrown in a landfill, washed into the ocean, or placed in a home compost bin.

bioplastics Plastics derived from natural, renewable biomass sources, such as corn starch, microbiota, or vegetable fats and oils. Bioplastic can be made from agricultural by-products and also from used plastic bottles and other containers using microorganisms. They are categorized as "other" resins, or #7 plastics. Not all bioplastics are biodegradable.

bottle bill Legislation that provides a monetary incentive to return beverage containers for recycling. Oregon was the first US state to pass such a law, in 1971. Today the 10 US states with bottle bills boast a 70 percent average recycling rate, compared with an overall rate in the United States of 33 percent. Also known as *container deposit laws.*

bricolage The improvisation of items from secondhand materials; to construct something by using whatever is available.

CE100 Stands for the *Circular Economy 100* at the Ellen MacArthur Foundation. A global platform bringing together leading companies, emerging innovators, and regions to accelerate the transition to a circular economy.

Circular Design Test A test for closed-loop recycling: a package or product is considered a winner if it can be circulated from production to consumption and back again, over and over (e.g., a glass bottle turned back into a new glass bottle).

circular economy The make-use-recycle-remanufacture concept wherein we keep resources in use for as long as possible, extract their maximum value while in use, and then recover and regenerate products and materials at the end of each service life to reintegrate them into the supply chain. The focus is on durability, recyclability, and use of renewable resources, including energy inputs.

closed-loop recycling A system wherein a package can be recycled back into itself; ideally, it can be circulated from production to consumption and back again repeatedly. *See also* Circular Design Test.

CO_2 Stands for *carbon dioxide,* a greenhouse gas.

compost A nutrient-rich mixture of decomposed organic substances that can be used to grow plants. Looks and smells like soil. Food waste and some paper products can be diverted from linear disposal and converted into compost.

compostable bioplastics A subset of biodegradable bioplastics intended to break down in a composting site at a rate consistent with other known compostable materials, such as paper, food waste, and yard trimmings. Like biodegradable bioplastics, most compostable bioplastics will break down only in a controlled, high-temperature industrial composting facility.

composting The process of converting organic materials such as food scraps into energy or compost, for at least a second life. Also known as *organics recycling*.

consumer packaged goods (CPG) Products that are sold quickly and at relatively low cost. Examples include nondurable goods such as packaged foods, beverages, toiletries, over-the-counter drugs, and other consumables. Also known as *fast-moving consumer goods (FMCG)*.

contaminants Any items that interfere with the conversion of recyclables into a high-value, marketable end-product. Nonorganic items that are typically unrecyclable—such as plastic bags, Styrofoam, and flexible packaging—taint the waste stream by diminishing the quality and value of processed material; nutrient contaminants, such as food waste and residues, are also a common problem in recycling.

corporate social responsibility Self-regulation a by business to account for the impacts of its activities on the economy, sociopolitical systems, and the environment, including the generation of packaging waste—operational, manufacturing, and post-consumer. The term was coined as a result of rising public concerns about ethical business practices.

CPG Stands for *consumer packaged goods*.

disposable Designed to be thrown away.

dual-use packaging Reusable containers that are structurally designed to serve a function after first product use; examples are condiments sold in decorative crocks and dry goods sold in reusable cloth bags or tin canisters.

durable goods Products that last for multiple uses and typically can be repaired when broken.

ECCC Stands for *Environment and Climate Change Canada,* the department within the government of Canada responsible for coordinating environmental policies and programs.

end-market demand The most important component of the recycling infrastructure; without a strong end market—producers and manufacturers that want to buy the material—recyclability of an item can be diminished to nonexistence.

energy recovery The process in which waste material is incinerated for the purpose of energy capture.

EPA Stands for *US Environmental Protection Agency,* the department within the government of the United States responsible for coordinating environmental policies and programs.

EPS Stands for *expanded polystyrene,* otherwise known as *Styrofoam.*

e-waste Used or discarded electronic accessories, devices, and products near or at the end of their useful lives. Mobile devices, landline telephones, laptops, and desktop computer parts fall into this category and very quickly become e-waste due to increasingly frequent technology upgrades. Unrecyclable municipally, many electronics contain heavy metals such as mercury, lead, and lithium that release into the environment when landfilled or incinerated. The EPA states that e-waste is the source of 70 percent of heavy metals in landfill.

extended producer responsibility (EPR) A policy approach under which producers are given a significant responsibility—financial, physical, or both—for the treatment or disposal of post-consumer waste. It extends a manufacturer's responsibility for reducing upstream product and packaging impacts to the downstream stage, when consumers are done with them. Unlike voluntary producer responsibility (VPR), EPR is mandated by law.

externalities An external effect, often unforeseen, as a result of the production or use of a product. For example, the current mass production of disposable plastic products and packaging coupled with a lack of recycling infrastructure results in plastic pollution.

fast fashion Clothing and apparel affordable enough to be considered disposable. Similar to packaging, today's material sourcing and production innovations in the fashion industry drive consumption by bringing down costs for manufacturers and consumers, allowing for a 52-season trend cycle, one for every week of the year. The goal of fast fashion is to motivate consumers to buy as many garments as possible, as quickly as possible, replacing and tossing cheaply made, trendy wardrobe items rather than reusing or repairing them.

feedstock Any raw, unprocessed material used to fuel a machine or industrial process to produce energy, goods, or finished products. In terms of new (virgin) materials, plastics and other synthetic materials are derived from petroleum (oil) feedstocks, glass from sand (silicon dioxide), paper from tree pulp, and aluminum from ore. Using secondary or "waste" materials as feedstocks offsets consumption of both finite and moderately renewable resources (e.g., forests).

flexible packaging In contrast with rigid packaging, flexible packaging is, as the name suggests, flexible. Common examples include bread bags, baby food pouches, and plastic shrink wraps. Composed of layers of plastic, paper, and/or metal, flexible packaging is considered difficult to recycle because the elements require separating at the material level. These iterations often include add-ons and closures that make the product convenient but which also require separating and are ultimately too small to recycle.

FMCG Stands for *fast-moving consumer goods.*

greenwashing A form of spin in which "green" marketing is deceptively used to promote the perception that an organization's products, aims, or policies are environmentally friendly when they are not.

gross domestic product (GDP) The sum of consumption, investment, government spending, and net exports; it is an indicator of the economic resources devoted by taxpayers to environmental protection.

HDPE Stands for *high-density polyethylene,* or #2 plastics.

input Energy and/or materials that go into a system to produce an output. In the modern recycling system, highly separated material inputs require less energy to process and are considered the most valuable. In nature all natural materials are valuable inputs in a regenerative system.

landfilling Burying waste between layers of earth or tossing it into a pile or hole; it is one of the most economically viable ways of disposing of waste in the current recycling infrastructure, as opposed to recycling or repurposing.

lightweighting A packaging trend wherein conventional packaging is replaced with a lighter-weight alternative and/or the overall amount of material used in packaging is reduced.

linear economy The current one-way, make-use-dispose economic system that views products and packaging as disposable after first use.

litter Random, undisposed trash—such as paper, cans, and bottles—that is left lying in an open or public place, prone to be eaten by animals, end up in oceans and waterways, or release chemicals into the environment.

materials recovery facility (MRF) The intermediate processing destination for potentially recyclable goods once they are collected from a home or business. It is a specialized facility that receives, separates, and prepares waste materials for sale to end-user manufacturing companies.

meaningful connections The desire to be closer to family, one's community, and the environment; a significant factor in defining a good life for today's consumer.

microplastic Tiny, often microscopic particles that form when plastics break down into smaller particle sizes in the environment.

MRF Pronounced "murf." Stands for *materials recovery facility;* also stands for *materials reclamation facility, materials recycling facility,* and *multi-reuse facility.*

multi-compositional Composed of several layers of different types of papers, plastics, and/or metal foils. These configurations include aseptic cartons and pouches, which create moisture barriers and protection from UV light, which is useful in food and beverage packaging. These are difficult to recycle because the components require separating at the material level. Also known as *multi-material hybrid packaging*.

multimaterial hybrid packaging Includes stand-up food pouches and aseptic cartons—multi-compositional containers that currently cannot be economically recycled. By combining the properties of different materials, multimaterial packaging can often offer enhanced performance versus its monomaterial alternatives, such as providing oxygen and moisture barriers at reduced weight and costs; represents about 13 percent of the market by weight. Also known as *multi-compositional packaging*.

NGO Stands for *nongovernmental organization*.

NIR Stands for *near-infrared;* a type of detection system.

nutrient-contaminated materials Ranging from dining disposables to coffee capsules, these are often difficult to sort and clean for high-quality recycling; the category includes applications and configurations that are prone to be mixed with organic contents during or after use.

OCC screen Stands for *old corrugated cardboard screen*. A means of separating materials in a materials recovery facility, it imparts a bouncing action on the materials stream; larger rigid objects like cardboard bounce over the top, and smaller materials drop through.

open-loop recycling A system wherein a material is recycled into other types of products (e.g., soda bottle into fiber).

output The opposite of input, the resulting yield produced by any system. In the recycling system, high-value outputs are those for which there is a demand in the market for secondary materials.

oxo-biodegradable plastics Plastics made from polymers such as polyethylene, polypropylene, and polystyrene containing extra ingredients (metal salts); they are designed to degrade and biodegrade in the open environment, triggered by UV radiation or heat. Instead of harmlessly being absorbed back into the environment, however, they simply rapidly break apart, turning into harmful microplastics.

packaging and printed paper (PPP) Representing 96 million metric tons of municipal solid waste generated each year, PPP includes steel, aluminum, glass, plastic, wood pallets, paper, and paperboard; about half of this amount is recycled.

participation One of four factors that affect the recycling of an item in the current infrastructure; a prerequisite for consumer participation in recycling is access to the system, as are education and engagement.

PCR Stands for *post-consumer recycled.*

PE Stands for *polyethylene;* includes both high-density polyethylene (HDPE, or #2 plastics) and low-density polyethylene (LDPE, or #4 plastics).

peer-to-peer businesses Companies that facilitate human connections, authentic experiences, and flexible work opportunities with new levels of searchability, personalization, transparency, and trust. *See also* sharing economy.

PET Stands for *polyethylene terephthalate,* or #1 plastics.

PLA Stands for *polylactic acid,* a transparent bioplastic produced from corn or dextrose.

planned obsolescence A policy of planning or designing a product with a limited useful life so that it will become obsolete, that is, unfashionable or no longer functional after a certain period of time.

plasticizers The molecules added to polymers that push the polymer molecules slightly farther apart, making the material softer and more flexible.

plastic packaging The throwaway material that represents the largest application of plastics in the economy, representing 26 percent of the total volume of plastics used; only 14 percent is collected for recycling.

point-of-sale (POS) material In-store signage, merchandising units, billboards, and the like, used to draw consumer attention in the retail setting; a type of pre-consumer waste.

post-consumer recycled (PCR) materials Sorted post-consumer materials such as plastics, paper and cardboard, and ferrous and nonferrous metal. Also known as *recycled raw materials (RRM)*.

post-consumer waste Waste that the consumer sees and interacts with and discards.

pouch A form of lightweighted packaging increasingly favored by food, pharmaceutical, medical, and beverage industries worldwide. Typically composed of multilayer films in combinations of plastic, paper, and aluminum, pouches bring down costs for producers, extend the shelf life of goods, and offer convenience and ease of use for consumers. While cited as an environmental positive for taking up less space in landfills, pouches are not recyclable municipally.

PP Stands for *polypropylene,* or #5 plastics; a thermoplastic polymer.

PPP Stands for *packaging and printed paper.*

pre-consumer waste Material produced during the manufacturing process that does not make it into the final product; also including secondary packaging, shipping, and point-of-sale material, this waste stream is rarely visible to the consumer.

product stewardship The requirement that the manufacturer's responsibility for its product extends to management of that product and its packaging when it hits the end of its life.

PS Stands for *polystyrene,* or #6 plastics.

PVC Stands for *polyvinyl chloride,* or #3 plastics; otherwise known as *vinyl.*

recommerce A repurposing model that connects buyers and sellers of secondhand goods; the great value is the ability to access and own gently used goods without the packaging and product waste associated with buying new.

recycle To repurpose a waste object by valuing the component materials from which it is made.

recycled content The percentage of recycled material in a new product or package.

recycled materials Materials that are collected, processed, and purchased for use back in new production.

recycled raw materials (RRM) Sorted post-consumer materials such as plastics, paper and cardboard, and ferrous and nonferrous metal. Also known as *post-consumer recycled (PCR) materials.*

recycling The conversion of discarded items into new, usable materials.

refillables Can refer to products or packaging consumers or retailers provide that can be refilled rather than tossed after one use; a bold alternative for many products sold in lightweighted, disposable packaging, such as cosmetics or household cleaners. Like buying in bulk, refilling is very much contingent on retailers willing to integrate the service.

regulated waste Items for which there are strict federal regulations controlling their handling, processing, and disposal. Often also referred to by the EPA as "universal waste," the main categories are batteries, pesticides, mercury-containing equipment, and lamps. Some states add their own items to the list, including (in limited situations) aerosol cans, antifreeze, lighting ballasts, barometers, cathode ray tubes, e-waste, oil-based finishes, paint, and pharmaceuticals.

rental and subscription model An alternative to ownership wherein customers rent, borrow, and subscribe to goods and services; it's the preference of consumers who desire more flexibility, more variety, and more abundance without the waste. *See also* sharing economy.

resin identification code The set of numbers and symbols appearing on plastic products that identify the plastic resin from which the product is made. The system was developed originally by the Society of the Plastics Industry (now the Plastics Industry Association) in 1988 but has been administered by ASTM International since 2008. Not to be mistaken for "recycling numbers," of which there is no such thing.

resource insecurity The state of not having enough resources to go around.

reuse *noun:* The creation or support of products that are durable and built to last, allowing consumers to use them repeatedly and for longer periods of time, offsetting the need for new materials and waste disposal with each cycle. Reuse vests products with value and keeps them at high utility.

reuse *verb:* To repurpose a waste object by valuing the material from which it is made, the form that the material is in, and the intention of the form; to use again.

reverse logistics The process of diverting usable material from linear disposal for the purpose of reusing it or capturing its value; necessitates systems that initially require resources to develop and execute.

rHDPE Stands for *recycled high-density polyethylene,* or *recycled HDPE.*

rPET Stands for *recycled polyethylene terephthalate,* or *recycled PET.*

rPP Stands for *recycled polypropylene,* or *recycled PP.*

sachet An individual portion pack; small, single-use, plastic pouch–like items that are inexpensive and popular for producers and consumers alike. They are used to package everything from condiments and soup starters to shampoo and instant drinks.

secondary packaging What bulk products come in, like a crate for soda or a box of boxes of cereal, intended to protect the primary packaging and deliver the product; a type of pre-consumer waste.

separation One of four factors that affect the recycling of an item in the current infrastructure; it describes the separation achieved at the point of collection, the technical aspects of sorting within a MRF and other reclamation facilities, and the resulting separation and quality of end product (recyclables).

sharing economy An economic model in which wealth and resources (goods and services) are accessed and consumed through sharing and borrowing rather than purchasing and owning. Modern examples include Lyft, Airbnb, and Rent the Runway.

single-stream recycling Introduced in the United States in 1995, a system wherein municipal solid waste is collected in a single bin rather than separated by material type; it increases consumer participation because it is easy to understand, but it requires added separation on the back end and diminishes the quality of the recyclables accepted.

single-use plastics Materials designed to be used only once and then thrown away or recycled. Designed this way for convenience and affordability, only 5 percent of the material value is retained, as these items are not recovered. Most plastic packaging falls into this category, including bottles, pouches, and sachets. Also known as *"disposable" plastics*.

small-format packaging Items such as sachets, tear-offs, lids, straws, candy wrappers, and small pots that have no economic reuse or recycling pathway; represents about 10 percent of the market by weight and 35 percent to 50 percent by number of items.

source reduction Using less material to produce products and packaging, a common resource and waste management practice for most businesses. Using less costs less, and a lighter, smaller package saves on material and transportation costs.

stewardship organization A nonprofit entity, approved by a government oversight agency, that works on behalf of producers to maximize operational efficiency. With regard to products and packaging, such organizations contract for collection and recycling services, conduct education and outreach, report to the government oversight body, and determine the fees that each company must pay to the organization.

storied plastics Resins collected by waste stream, sorted by material type, and traced back to the point of origin (e.g., recycled ocean plastic); they can help brands circumvent some of the prohibitive barriers of cost, quality, and diminished recyclability for mixed recycled resins with supply-chain security and added marketing value.

take-back programs A program to "take back" and collect discarded items not accepted curbside. Operating outside of public, municipal programs and often hosted by manufacturers, retailers, and brands, take-back programs are an alternative to throwing items directly in the trash and potential non-compliance with federal regulations, but they are not always for the purpose of recycling. *See also* regulated waste.

tear-offs A packaging configuration that requires the consumer to "tear off" a piece of it in order to open it and use the consumable product inside. These include sachets, pouches, and envelopes with tear-off corners or tops, as well as containers with peel-off plastic films (e.g., the top of a potato chip can and the plastic wrap on a frozen meal).

true cost accounting An economic principle that takes into account all the external costs factored out of the cost of a product; it helps consumers understand the real cost of the products they buy, which may include environmental and ethical externalities as well as financial costs in terms of the taxes they pay to dispose of and recycle the product within the public system.

uncommon plastic Materials such as PVC (polyvinyl chloride), PS (polystyrene), and EPS (expanded polystyrene, aka Styrofoam) categorized as #7 "other" plastics. While often technically recyclable, these plastics are not economically viable to sort and recycle because their small volume prevents effective economies of scale; represents about 10 percent of the market by weight.

UV Stands for *ultraviolet*. Referring to naturally occurring electromagnetic radiation present in sunlight, UV rays can damage plastic with long-term exposure.

virgin Newly manufactured material not yet used in production. The opposite of recycled or secondary raw material, virgin material has been the preferred choice of plastics, paper, and even aluminum and glass with regard to price, function, aesthetics, and supply, as the current infrastructure deems them the most economically viable. Demand for virgin material drains finite resources in lieu of reintegrating usable (recycled) material into new production.

voluntary producer responsibility (VPR) program A form of product stewardship that extends the manufacturer's responsibility for its product to post-consumer management of that product and its packaging. Unlike extended producer responsibility (EPR), VPR is not mandated by law.

waste stream A category of waste, such as plastic cutlery or coffee cups.

Zero Waste Box A consumer recycling solution for hundreds of difficult-to-recycle waste streams. The boxes are turnkey in that they are ready to use upon delivery and contain a prepaid shipping label. Consumers select the desired size and category, fill with the accepted waste streams, and send to TerraCycle for processing. Once collected, the items are recycled for use in new products.

zero-waste-to-landfill (ZWTL) A business concept in which solid waste produced at respective facilities is not landfilled but instead is reused, recycled, composted, or disposed of via some other outlet. Some companies regard ZWTL as a guiding ideal rather than a benchmark, and many seek certification for visibility.

ZMWTL Stands for *zero-manufacturing-waste-to-landfill*.

ZNHWTL Stands for *zero-nonhazardous-waste-to-landfill*.

ZWTL Stands for *zero-waste-to-landfill*.

Notes

FOREWORD

1. Fred Pearce, "From Ocean to Ozone: Earth's Nine Life-Support Systems," *New Scientist,* February 27, 2010, https://www.newscientist.com/article/dn18574-earths-nine-life-support-systems-biodiversity (accessed May 5, 2018).

2. United Nations, "World Population Projected to Reach 9.8 Billion in 2050, and 11.2 Billion in 2100" (news release), June 21, 2017, https://www.un.org/development/desa/en/news/population/world-population-prospects-2017.html (accessed May 5, 2018).

3. Ellen MacArthur Foundation, World Economic Forum, and McKinsey & Company, "The New Plastics Economy: Rethinking the Future of Plastics," January 2016, http://www3.weforum.org/docs/WEF_The_New_Plastics_Economy.pdf (accessed May 23, 2018).

INTRODUCTION

From Linear to Circular

1. Ellen MacArthur Foundation, World Economic Forum, and McKinsey & Company, "The New Plastics Economy: Rethinking the Future of Plastics," January 2016, http://www3.weforum.org/docs/WEF_The_New_Plastics_Economy.pdf (accessed May 5, 2018).

2. Chris Tyree and Dan Morrison, "Invisibles: The Plastic Inside Us," *Orb,* https://orbmedia.org/stories/Invisibles_plastics/multimedia (accessed May 5, 2018).

3. Susan Smillie, "From Sea to Plate: How Plastic Got into Our Fish," *The Guardian,* February 14, 2017, https://www.theguardian.com/lifeandstyle/2017/feb/14/sea-to-plate-plastic-got-into-fish (accessed May 5, 2018).

4. US Centers for Disease Control and Prevention, "Expected New Cancer Cases and Deaths in 2020," https://www.cdc.gov/cancer/dcpc /research/articles/cancer_2020.htm (accessed May 5, 2018).

5. Center for International Environmental Law, "Fueling Plastics: Plastic Industry Awareness of the Ocean Plastics Problem," http://www.ciel .org/wp-content/uploads/2017/09/Fueling-Plastics-Plastic-Industry -Awareness-of-the-Ocean-Plastics-Problem.pdf (accessed May 5, 2018).

6. US Environmental Protection Agency, "Municipal Solid Waste," https://archive.epa.gov/epawaste/nonhaz/municipal/web/html (accessed May 5, 2018).

7. Laura Parker, "A Whopping 91% of Plastic Isn't Recycled," *National Geographic,* July 19, 2017, https://news.nationalgeographic.com /2017/07/plastic-produced-recycling-waste-ocean-trash-debris -environment (accessed May 5, 2018).

8. TerraCycle, https://www.terracycle.com (accessed May 5, 2018).

CHAPTER 1

Plastic, Packaging, and the Linear Economy

1. Anthony L. Andrady and Mike A. Neal, "Applications and Societal Benefits of Plastics," *Philosophical Transactions of the Royal Society* 364, no. 1526 (2009): 1977–84; doi: 10.1098/rstb.2008.0304 (accessed May 5, 2018).

2. Christel Trimborn, "Galalith—Jewelry Milk Stone," Ganoskin.com, https://www.ganoksin.com/article/galalith-jewelry-milk-stone (accessed May 5, 2018).

3. American Petroleum Institute, "Power Past Impossible," published February 5, 2017, https://www.youtube.com/watch?v=9iw4fcPezn4 (accessed May 14, 2018).

4. Andrady and Neal, "Applications and Societal Benefits."

5. American Chemical Council, "How Plastics Are Made," https:// plastics.americanchemistry.com/How-Plastics-Are-Made (accessed May 5, 2018).

6. Clive Everton, *The History of Snooker and Billiards* (Colchester, UK: The Book Service, 1986), 11.

7. Susan Freinkel, *Plastic: A Toxic Love Story* (New York: Houghton Mifflin Harcourt, 2011).

8. *Landmarks of the Plastics Industry, 1862–1962* (Welwyn Garden City, UK: Imperial Chemical Industries [Plastics Division], 1962), 13–25.

9. Heather Rogers, "A Brief History of Plastic," *Brooklyn Rail*, May 1, 2005, https://brooklynrail.org/2005/05/express/a-brief-history-of -plastic (accessed May 5, 2018).

10. Rogers, "Brief History of Plastic."

11. Tom Szaky, "Creating the World's First Recyclable Shampoo Bottle Made with Beach Plastic," HuffPost, January 19, 2017, https://www .huffingtonpost.com/entry/creating-the-worlds-first_b_14266776 .html (accessed May 5, 2018).

12. Gordon L. Robertson, *Food Packaging: Principles and Practice*, 2nd ed. (Boca Raton, FL: CRC Press, 2005).

13. Science History Institute, "The History and Future of Plastics," https:// www.sciencehistory.org/the-history-and-future-of-plastics (accessed May 5, 2018).

14. Statista, "Production of Polyethylene Terephthalate Bottles World- wide from 2004 to 2021 (in Billions)," https://www.statista.com /statistics/723191/production-of-polyethylene-terephthalate-bottles -worldwide (accessed May 5, 2018).

15. Sandra Laville and Matthew Taylor, "A Million Bottles a Minute: World's Plastic Binge 'As Dangerous as Climate Change,'" *The Guardian*, June 28, 2017, https://www.theguardian.com/environment/2017/jun/28 /a-million-a-minute-worlds-plastic-bottle-binge-as-dangerous-as -climate-change (accessed May 23, 2018).

16. Joe Iles, "Five Ways to End Plastic Waste," *Circulate*, March 1, 2018, http://circulatenews.org/2018/03/five-ways-end-plastic-waste (accessed May 5, 2018).

CHAPTER 2

Where Did Public Recycling Come From, and Where Is It Going?

1. "The History of Paper," Silkroad Foundation, http://www.silk-road .com/artl/papermaking.shtml (accessed May 8, 2018).

2. Teresa Gowan, "American Untouchables: Homeless Scavengers in San Francisco's Underground Economy," *International Journal of Sociology*

and Social Policy 17, no. 3/4 (1997): 159–90; doi: 10.1108/eb013304 (accessed May 8, 2018).

3. Martin Medina, *The World's Scavengers: Salvaging for Sustainable Consumption and Production* (New York: AltaMira Press, 2007).

4. Jacqueline Foertsch, *American Culture in the 1940s* (Edinburgh, Scotland: Edinburgh University Press, 2008), xvi.

5. "Asks Scraps to Win War," *New York Times,* January 30, 1942, https://www.nytimes.com/1942/01/30/archives/asks-scraps-to-win-war-salvage-for-victory-drive-stresses.html (accessed May 8, 2018).

6. Wikipedia, s.v. "Paper Salvage 1939–1950" (citing the National Archives, INF 2/70, "Central Office of Information: Domestic Fuel Economy & Waste Paper Salvage Campaigns"), last modified October 17, 2017, 12:49, https://en.wikipedia.org/wiki/Paper_Salvage_1939%E2%80%9350#cite_ref-3.

7. Tim Cooper, "Challenging the 'Refuse Revolution': War, Waste and the Rediscovery of Recycling, 1900–1950" (research paper University of St. Andrews, July 19, 2007), https://ore.exeter.ac.uk/repository/bitstream/handle/10036/28893/Microsoft%20Word%20-%20Recycling%20Pre-cceptance.pdf?sequence=2 (accessed May 8, 2018).

8. Cooper, "Challenging the Refuse Revolution."

9. William M. Rohe and Harry L. Watson, *Chasing the American Dream: New Perspectives on Affordable Homeownership* (Ithaca, NY: Cornell University Press, 2007).

10. Lucy R. Lippard, "New York Comes Clean: The Controversial Story of the Fresh Kills Dumpsite," *The Guardian*, October 28, 2016, https://www.theguardian.com/cities/2016/oct/28/new-york-comes-clean-fresh-kills-staten-island-notorious-dumpsite, (accessed May 8, 2018).

11. Earth Day Network, "The History of Earth Day," https://www.earthday.org/about/the-history-of-earth-day (accessed May 8, 2018).

12. Finis Dunaway, *Seeing Green: The Use and Abuse of American Environmental Images* (Chicago: University of Chicago Press, 2015).

13. Kat Eschner, "How the 1970s Created Recycling as We Know It," Smithsonian.com, November 15, 2017, https://www.smithsonianmag.com/smart-news/how-1970s-created-recycling-we-know-it-180967179/#8XMsO5XRSq3JzvvO.99 (accessed May 8, 2018).

14. US Environmental Protection Agency, "Municipal Solid Waste Generation, Recycling, and Disposal in the United States: Facts and Figures for 2012," September 2015, https://www.epa.gov/sites/production/files/2015-09/documents/2012_msw_fs.pdf (accessed May 8, 2018).

15. Bartow J. Elmore, *Citizen Coke: The Making of Coca-Cola Capitalism* (New York: W. W. Norton, 2015), 233.

16. Philip Shabecoff, "With No Room at the Dump, US Faces a Garbage Crisis," *New York Times,* June 29, 1987, http://www.nytimes.com/1987/06/29/us/with-no-room-at-the-dump-us-faces-a-garbage-crisis.html (accessed May 8, 2018).

17. Joseph Kennedy, "Recycling through the Ages: 1980s," Plastic Expert, August 4, 2014, http://www.plasticexpert.co.uk/recycling-ages-1980s (accessed May 8, 2018).

18. David Vogel, "The Revival of Corporate Social Responsibility," in *The Market for Virtue: The Potential and Limits of Corporate Social Responsibility* (Washington, DC: Brookings Institution, 2006); available at https://www.brookings.edu/wp-content/uploads/2016/07/marketforvirtue_chapter.pdf (accessed May 14, 2018).

19. Jo Thomas, "After Growing in Success, Recycling Faces Obstacles," *New York Times*, November 17, 1994, http://www.nytimes.com/1994/11/27/nyregion/after-growing-in-success-recycling-faces-obstacles.html?pagewanted=all (accessed May 8, 2018).

20. "The Truth about Recycling," *The Economist*, June 7, 2007, http://www.economist.com/node/9249262 (accessed May 8, 2018).

21. Mary H. Cooper, "The Economics of Recycling," *CQ Researcher* 8, no. 12 (1998); http://library.cqpress.com/cqresearcher/document.php?id=cqresrre1998032700 (accessed May 8, 2018).

22. Brenda Platt and Neil Seldman, "Wasting and Recycling in the United States 2000," GrassRoots Recycling Network, 2000, http://www.grrn.org/assets/pdfs/wasting/WRUS.pdf (accessed May 8, 2018).

23. US Environmental Protection Agency, "Municipal Solid Waste," https://archive.epa.gov/epawaste/nonhaz/municipal/web/html (accessed May 8, 2018).

24. "Great Pacific Garbage Patch: Pacific Trash Vortex," NationalGeographic.com, www.nationalgeographic.org/encyclopedia/great-pacific-garbage-patch (accessed May 8, 2018).

25. Congressman Frank Pallone, "Legislation Requiring Clean Up of Rail Solid Waste Sites Becomes Law" (news release), October 20, 2008, https://pallone.house.gov/press-release/legislation-requiring-clean -rail-solid-waste-sites-becomes-law (accessed May 8, 2018).

CHAPTER 3

The State of the Recycling Industry

1. Frances Bula, "China's Tough New Recycling Standards Leaving Canadian Municipalities in a Bind," *The Globe and Mail,* January 8, 2018, https://www.theglobeandmail.com/news/national/chinese-ban-on -foreign-recyclables-leaving-some-canadian-cities-in-the-lurch /article37536117 (accessed May 8, 2018).

2. Daniel Hoornweg, Perinaz Bhada-Tata, and Chris Kennedy, "Environment: Waste Production Must Peak This Century," *Nature,* October 30, 2013, https://www.nature.com/news/environment-waste-production -must-peak-this-century-1.14032 (accessed May 8, 2018).

3. Agence de l'Environnement et de la Maîtrise de l'Énergie, "Waste Prevention in France: 2011 Status Report," 2012, http://www.ademe.fr /sites/default/files/assets/documents/85945_7505_preventiondechets _en.pdf (accessed May 8, 2018).

4. US Census Bureau, "US and World Population Clock," https://www .census.gov/popclock (accessed May 8, 2018).

5. Daniel Hoornweg and Perinaz Bhada-Tata, "What a Waste: A Global Review of Solid Waste Management," World Bank urban development series, knowledge paper no. 15, March 2012, https://openknowledge .worldbank.org/handle/10986/17388, 83 (accessed May 8, 2018).

6. US Environmental Protection Agency, "Advancing Sustainable Materials Management: 2014 Fact Sheet," November 2016, https://www .epa.gov/sites/production/files/2016-11/documents/2014_smmfact sheet_508.pdf, 10–11 (accessed May 8, 2018).

7. US EPA, "Advancing Sustainable Materials Management."

8. US EPA, "Advancing Sustainable Materials Management."

9. Ted Michaels and Ida Shiang, "Energy Recovery Council: 2016 Directory of Waste-to-Energy Facilities," May 2016, http://energyrecovery council.org/wp-content/uploads/2016/05/ERC-2016-directory.pdf, 3 (accessed May 8, 2018).

10. US Environmental Protection Agency, "Summary of the Resource Conservation and Recovery Act," last modified August 2017, https://www.epa.gov/laws-regulations/summary-resource-conservation -and-recovery-act (accessed May 8, 2018).

11. Matt Richtel, "San Francisco, 'the Silicon Valley of Recycling,'" *New York Times*, March 25, 2016, https://www.nytimes.com/2016/03/29 /science/san-francisco-the-silicon-valley-of-recycling.html (accessed May 8, 2018).

12. "What Is the 'Fantastic Three' Program?" SF Environment, https:// sfenvironment.org/zero-waste-faqs#fantastic-three (accessed May 8, 2018).

13. "Transparency: Environment and Climate Change Canada," Government of Canada, last modified November 2, 2017, https://www .canada.ca/en/environment-climate-change/corporate/transparency .html (accessed May 8, 2018).

14. Nicole Martin, "City Trying to Get Torontonians to Stop Filling Recycling Bins with Garbage," CBC, June 1, 2017, http://www.cbc .ca/news/canada/toronto/recycling-toronto-garbage-1.4141579 (accessed May 8, 2018).

15. Avani Babooram and Jennie Wang, "Recycling in Canada," Statistics Canada, December 9, 2013, http://www.statcan.gc.ca/pub/16 -002-x/2007001/article/10174-eng.htm (accessed May 9, 2018).

16. "Mexico Is the World Leader for PET Recycling," Geo-Mexico, April 11, 2016, http://geo-mexico.com/?p=13773 (accessed May 9, 2018).

17. PROtrash, "Mexico's Massive 24 Billion Dollar Recycling Opportunity," HuffPost, August 27, 2017, https://www.huffingtonpost.com /protrash/mexicos-massive-24-billio_b_11730898.html (accessed May 9, 2018).

18. International Trade Administration, "Mexico—Plastics and Resins," September 19, 2017, https://www.export.gov/article?id=Mexico -Plastic-Materials-Resins (accessed May 9, 2018).

19. "Environmental Protection Expenditure," Eurostat, http://ec.europa.eu /eurostat/web/environment/environmental-protection-expenditure (accessed May 9, 2018).

20. J. Jara-Samaniego, M. D. Pérez-Murcia, M. A. Bustamante, et al., "Development of Organic Fertilizers from Food Market Waste and

Urban Gardening by Composting in Ecuador," *PLoS ONE* 12, no. 7 (2017); http://journals.plos.org/plosone/article?id=10.1371/journal .pone.0181621 (accessed May 9, 2018).

21. SUEZ Environnement, "SUEZ Environnement, through Its Subsidiary Sita Atlas, Wins 20-Year Contract Worth €90 Million to Build and Operate the Waste Elimination and Recycling Facility in Meknes" (news release), February 7, 2014, https://www.vfb.be/vfb /Media/Default/archief_pers/suez-environnement/SUEZ%20 ENVIRONNEMENT%20-%20SITA%20ATLAS%20wins%2020 -year%20contract%20worth%20%E2%82%AC90%20mllion%20 (10.2.2014).pdf (accessed May 16, 2018).

22. James Clasper, "How the Moroccan City of Meknes Got Its Sprawling Landfill under Control," Virgin.com, October 30, 2017, https://www .virgin.com/virgin-unite/how-moroccan-city-meknes-got-its-sprawl ing-landfill-under-control (accessed May 16, 2018).

23. European Commission, "Circular Economy Package: Questions & Answers" (news release), December 2, 2015, http://europa.eu/rapid /press-release_MEMO-15-6204_en.htm (accessed May 9, 2018).

24. "Municipal Waste Statistics," Eurostat, last modified April 24, 2018, 18:41, http://ec.europa.eu/eurostat/statistics-explained/index.php /Municipal_waste_statistics (accessed May 9, 2018).

25. Newcastle-under-Lyme Borough Council, "Recycling and Waste: Your Collection Services," https://www.newcastle-staffs.gov.uk /all-services/recycling-and-waste/your-collection-services (accessed May 9, 2018).

26. Nordic Council of Ministers, "Nordic Improvements in Collection and Recycling of Plastic Waste," 2015, http://norden.diva-portal.org /smash/get/diva2:788308/FULLTEXT01.pdf (accessed May 9, 2018).

27. Jonas Fredén, "The Swedish Recycling Revolution," Swedish Institute, March 29, 2017, https://sweden.se/nature/the-swedish-recycling -revolution (accessed May 9, 2018).

28. Vilhelm Carlström, "Sweden Is So Good at Recycling That It's Now Getting a Deposit-Refund System for Plastic Bags," Business Insider Nordic, May 5, 2017, http://nordic.businessinsider.com/sweden-is -so-good-at-recycling-its-now-getting-a-deposit-refund-system-for -plastic-bags-2017-5 (accessed May 9, 2018).

CHAPTER 4

Who Is Responsible for Recycling Packaging?

1. Federal Trade Commission, "Fair Packaging Labeling Act," https://www.ftc.gov/enforcement/rules/rulemaking-regulatory-reform-proceedings/fair-packaging-labeling-act (accessed May 9, 2018).

2. US Food and Drug Administration, "Guidance for Industry: Food Labeling Guide," revised January 2013, https://www.fda.gov/Food/GuidanceRegulation/GuidanceDocumentsRegulatoryInformation/LabelingNutrition/ucm2006828.htm (accessed August 1, 2018).

3. Matt Prindiville and Jamie Rhodes, "5 Reasons EPR Is the Answer for Plastics Recycling," Sustainable Brands, May 19, 2016, http://www.sustainablebrands.com/news_and_views/packaging/matt_prindiville/why_epr_answer_plastics_recycling (accessed May 9, 2018).

CHAPTER 5

Recycled versus Recyclable

1. Association of Plastic Recyclers, "The APR Design® Guide for Plastics Recyclability," http://www.plasticsrecycling.org/apr-design-guide/apr-design-guide-home (accessed May 9, 2018).

2. US Environmental Protection Agency, "Sustainable Materials Management: Non-Hazardous Materials and Waste Management Hierarchy," https://www.epa.gov/smm/sustainable-materials-management-non-hazardous-materials-and-waste-management-hierarchy (accessed May 9, 2018).

3. PureCycle Technologies, "PureCycle Technologies and P&G Introduce Technology That Enables Recycled Plastic to Be Nearly-New Quality," https://www.prnewswire.com/news-releases/purecycle-technologies-and-pg-introduce-technology-that-enables-recycled-plastic-to-be-nearly-new-quality-300491368.html (accessed May 9, 2018).

4. Materials Recovery for the Future, https://www.materialsrecoveryforthefuture.com (accessed May 19, 2018).

5. The Recycling Partnership, https://recyclingpartnership.org (accessed May 19, 2018).

CHAPTER 6

Designing Packaging for the Simple Recycler: How MRFs Work

1. Recyclebank, "MRF [Materials Recovery Facility]," https://myrecycling .recyclebank.com/eco-library/mrf-materials-recovery-facility (accessed May 10, 2018).

2. Alexander J. Dubanowitz, "Design of a Materials Recovery Facility (MRF) for Processing the Recyclable Materials of New York City's Municipal Solid Waste," May 2000, http://www.seas.columbia.edu /earth/dubanmrf.pdf (accessed May 10, 2018).

3. Megan Willett, "Garbage Collectors Share the Most Bizarre Things They've Ever Found on the Job," Business Insider, December 17, 2014, http://www.businessinsider.com/garbage-collectors-find-bizarre -things-2014-12 (accessed May 10, 2018).

4. Devon Contract Waste, "What Happens at a Materials Recycling Facility (MRF)? How Does Our Sorting Machine Work?" published August 27, 2013, https://www.youtube.com/watch?v=SIVKmwzWSuc (accessed May 10, 2018).

5. Jennifer Kite-Powell, "This Recycling Robot Uses Artificial Intelligence to Sort Your Recyclables," Forbes, April 4, 2017, https://www .forbes.com/sites/jenniferhicks/2017/04/04/this-recycling-robot -uses-artificial-intelligence-to-sort-your-recyclables/#14fe34aa2d35 (accessed May 10, 2018).

CHAPTER 7

The Myth of Biodegradability

1. European Bioplastics, "Global Bioplastics Production Capacities Continue to Grow Despite Low Oil Price," November 30, 2016, http:// www.european-bioplastics.org/market-data-update-2016 (accessed May 10, 2018).

2. Smithers Pira, "Global Bioplastics for Packaging Market Forecast to Grow by 17% CAGR to 2022," April 2017, https://www.smitherspira .com/news/2017/april/global-bioplastics-for-packaging-market -forecasts (accessed May 10, 2018).

3. Renee Cho, "The Truth about Bioplastics," Earth Institute, December 14, 2017, https://phys.org/news/2017-12-truth-bioplastics.html (accessed May 10, 2018).

4. BusinessGreen Staff, "Unilever, PepsiCo Wash Hands of 'Biodegradable' Plastic," GreenBiz, November 7, 2017, https://www.greenbiz.com/article/unilever-pepsico-wash-hands-biodegradable-plastic (accessed May 10, 2018).

5. "Compostable," Compostable Info, http://www.compostable.info/compostable.htm (accessed May 10, 2018).

6. In a 1992 interview, archaeologist William J. Rathje recalled an order of guacamole he recently unearthed: "Almost as good as new, it sat next to a newspaper apparently thrown out the same day. The date was 1967." William Grimes, "Seeking the Truth in Refuse," *New York Times,* August 13, 1992, https://www.nytimes.com/1992/08/13/nyregion/seeking-the-truth-in-refuse.html (accessed May 10, 2018).

7. United Nations Environment Programme, "Biodegradable Plastics Are Not the Answer to Reducing Marine Litter, Says UN" (news release), November 17, 2015, https://www.unenvironment.org/news-and-stories/press-release/biodegradable-plastics-are-not-answer-reducing-marine-litter-says-un (accessed May 10, 2018).

8. Mary Catherine O'Connor, "Compostable or Recyclable? Why Bioplastics Are Causing an Environmental Headache," AlterNet, July 6, 2011, https://www.alternet.org/story/151543/compostable_or_recyclable_why_bioplastics_are_causing_an_environmental_headache (accessed May 10, 2018).

9. Plastics Industry Trade Association, Bioplastics Division, "BioPlastics Simplified: Attributes of Biobased and Biodegradable Plastics," February 2016, http://www.plasticsindustry.org/sites/plastics.dev/files/Bioplastics%20Simplified_0.pdf (accessed May 10, 2018).

10. Aditya Chakrabortty, "Secret Report: Biofuel Caused Food Crisis," *The Guardian,* July 3, 2008, https://www.theguardian.com/environment/2008/jul/03/biofuels.renewableenergy (accessed May 10, 2018).

11. Bioplastic Recycling, https://www.bioplasticrecycling.com (accessed May 10, 2018).

12. European Bioplastics, "Mechanical Recycling," http://www.european-bioplastics.org/bioplastics/waste-management/recycling (accessed May 10, 2018).

13. Coca-Cola, "Coca-Cola Produces World's First PET Bottle Made Entirely from Plants" (news release), June 3, 2015, http://www.coca-colacompany.com/press-center/press-releases/coca-cola

-produces-worlds-first-pet-bottle-made-entirely-from-plants (accessed May 10, 2018).

14. Renmatix, "Renmatix Secures $14M Investment from Bill Gates and Total, the Global Energy Major, in Concert with Signing of 1 Million Ton Cellulosic Sugar License" (news release), September 15, 2016, http://renmatix.com/products/announcements/renmatix-secures -14m-investment-from-bill-gates-and-total-the-global-energy -major-in-concert-with-signing-of-1-million-ton-cellulosic-sugar -license (accessed May 10, 2018).

CHAPTER **8**

Less Isn't Always More

1. Rogerio Hirose, Renata Maia, Anne Martinez, and Alexander Thiel, "Three Myths about Growth in Consumer Packaged Goods," McKinsey & Company, June 2015, https://www.mckinsey.com /industries/consumer-packaged-goods/our-insights/three-myths -about-growth-in-consumer-packaged-goods (accessed May 10, 2018).

2. "About the Carton Council," https://www.recyclecartons.com/about (accessed May 10, 2018).

3. Becci Vallis, "Why It's Time to Start Being More Eco-conscious When Choosing Your Beauty Products," *The Telegraph*, April 11, 2018, https://www.telegraph.co.uk/beauty/tips-tutorials/time-start -eco-conscious-choosing-beauty-products (accessed June 4, 2018).

4. "Thirst Inspiration," LIFEWTR, https://www.lifewtr.com (accessed May 10, 2018).

5. Kathryn Sukalich, "10 Things You Can Recycle through the Mail," Earth 911, March 13, 2013, https://earth911.com/home-garden /mail-back-recycling-programs (accessed May 10, 2018).

6. "Zero Waste Box," TerraCycle, https://zerowasteboxes.terracycle.com (accessed May 10, 2018).

7. Tom Szaky, "The Trend of Recycling Cosmetics Packaging," HuffPost, July 13, 2016 (updated July 14, 2017), https://www.huffingtonpost .com/entry/the-trend-of-recycling-co_b_10970210.html (accessed May 10, 2018).

8. Smithers Pira, "Packaging Material Outlooks—Towards a $1 Trillion Milestone in 2020," February 2016, https://www.smitherspira

.com/resources/2016/february/global-packaging-material-outlooks (accessed May 10, 2018).

9. Eric Fish, "Flexible Trends: Moving from Rigid to Flexible," Digital BNP Media, http://digital.bnpmedia.com/publication/?i=260669& article_id=2022182&view=articleBrowser&ver=html5#{"issue _id":260669,"view":"articleBrowser","article_id":"2022182"} (accessed May 10, 2018).

10. Eriko Arita, "Japan's 'Pouch Curry' Turns a Tasty 40," *Japan Times,* February 17, 2008, https://www.japantimes.co.jp/life/2008/02 /17/general/japans-pouch-curry-turns-a-tasty-40/#.V-KxevkrJ1v (accessed May 10, 2018).

11. Technavio, "Global Coffee Pods Market 2017–2021," May 2017, https://www.technavio.com/report/global-food-global-coffee -pods-market-2017-2021 (accessed September 18, 2018).

12. Nespresso, "The Positive Cup," https://www.nespresso.com/positive /us/en#!/sustainability (accessed May 10, 2018).

13. Michael J. Coren, "Soon You'll Be Able to Get Your Tortillas and Pet Food from Keurig-Like Pods," Quartz, May 3, 2016, https:// qz.com/674609/pods-are-turning-food-into-the-most-profitable -business-since-software (accessed May 10, 2018).

14. Technavio, "Global Flexible Packaging Market for Food and Beverages 2018–2022," May 2018, https://www.technavio.com/report /global-flexible-packaging-market-for-food-and-beverages -analysis-share-2018 (accessed September 18, 2018).

CHAPTER **9**

But More Isn't Always Better

1. Tom Szaky, "Rethinking Premium Packaging from a Recycling Viewpoint," *Packaging Digest,* October 2, 2017, http://www.packaging digest.com/sustainable-packaging/rethinking-premium-packaging -from-a-recycling-viewpoint-2017-10-02 (accessed May 21, 2018).

2. Jared Paben, "Meal Kits Not Sitting Well in Plastics Recovery Chain," Resource Recycling, July 5, 2017, https://resource-recycling.com /recycling/2017/07/05/meal-kits-not-sitting-well-in-plastics-recovery -chain (accessed May 10, 2018).

3. Lisa McTigue Pierce, "Does This Often-Overlooked Sustainable Packaging Metric Deserve More Attention?" *Packaging Digest,*

January 29, 2016, http://www.packagingdigest.com/sustainable
-packaging/does-this-often-overlooked-sustainable-packaging
-metric-deserve-more-attention-2016-01-29 (accessed May 10, 2018).

4. Editors of *The ULS Report*, "A Study of Packaging Efficiency as It
   Relates to Waste Prevention," January 2016, http://use-less-stuff
   .com/wp-content/uploads/2017/10//2016-Packaging-Efficiency
   -Study-1.19.16.pdf (accessed May 10, 2018). Also see http://use-less
   -stuff.com.

5. Lisa McTigue Pierce, "5 Ways to Prevent Packaging Waste," *Packaging
   Digest*, January 29, 2016, http://www.packagingdigest.com/sustainable
   -packaging/5-ways-to-prevent-packaging-waste-2016-01-29
   (accessed May 21, 2018).

6. Stress Engineering Services Inc., "Packaging & Packaging Lines,"
   http://www.stress.com/capabilities/packaging-packaging-lines
   (accessed May 10, 2018).

7. Esko, https://www.esko.com/en (accessed May 10, 2018).

8. SpecPage, https://www.specpage.com (accessed May 10, 2018).

9. Lisa McTigue Pierce, "Nature's Package Keeps Coconut Water Protected," *Packaging Digest*, June 19, 2017, http://www.packagingdigest
   .com/beverage-packaging/natures-package-keeps-coconut
   -water-protected-2017-06-19 (accessed May 21, 2018).

10. Rick Lingle, "High-Tech Cartons Converted with Distinction," *Packag-
    ing Digest*, February 2, 2015, http://www.packagingdigest.com/cartons
    /high-tech-converted-cartons-converted-with-distinction150202
    (accessed May 10, 2018).

11. Rick Lingle, "New Tech Reshapes Aluminum Beverage Bottles,"
    Packaging Digest, July 18, 2017, http://www.packagingdigest.com
    /beverage-packaging/new-tech-reshapes-aluminum-beverage
    -bottles1707 (accessed May 10, 2018).

12. Kate Bertrand Connolly, "Sustainable Packaging for Disney's Moana
    Doll Is Both Responsible and Fun," *Packaging Digest,* November 14,
    2016, http://www.packagingdigest.com/packaging-design/sustainable
    -packaging-for-disneys-moana-doll-is-both-responsible-and-fun
    -2016-11-14 (accessed May 10, 2018).

13. "Digital Printing," *Packaging Digest,* http://www.packagingdigest
    .com/digital-printing (accessed May 10, 2018).

14. Rick Lingle, "AR and VR in Packaging: Beyond the Buzz," *Packaging Digest,* July 26, 2017, http://www.packagingdigest.com/packaging -design/arvr-packaging-beyond-the-buzz1707 (accessed May 10, 2018).

15. Rick Lingle, "Augmented Reality App Complements Egg Carton Redesign," *Packaging Digest,* May 15, 2017, http://www.packagingdigest .com/packaging-design/ar-app-egg-carton-design-1705 (accessed May 21, 2018).

16. Lisa McTigue Pierce, "Packaging Design for Ecommerce Can Break the Rules," *Packaging Digest,* June 21, 2016, http://www.packagingdigest .com/packaging-design/packaging-design-for-ecommerce-can -break-the-rules-2016-06-21 (accessed May 10, 2018).

17. Lisa McTigue Pierce, "Amazon on Creating Ecommerce Packaging That's Great for All: Customers, Companies and the Environment," *Packaging Digest,* April 14, 2017, http://www.packagingdigest .com/optimization/amazon-on-creating-ecommerce-packaging -thats-great-for-customers-companies-and-environment-2017-04-14 (accessed May 21, 2018).

18. "Packaging and Prep Requirements," Amazon Seller Central, https:// sellercentral.amazon.com/gp/help/external/200141500?language =en-US&ref=mpbc_200243250_cont_200141500 (accessed May 21, 2018).

19. McTigue Pierce, "Amazon on Creating Ecommerce Packaging."

20. "Amazon Packaging Certification Case Studies," http://www.metaphase .com/wp-content/uploads/2017/07/Amazon_Packaging_Certifi cation_Case-Studies-v1.pdf (accessed September 18, 2018).

21. Kyla Fisher and Bob Lilienfeld, "Optimizing Packaging for an E-commerce World," January 2017, American Institute for Packaging and the Environment, https://c.ymcdn.com/sites/www.ameripen .org/resource/resmgr/PDFs/White-Paper-Optimizing-Packa.pdf (accessed May 21, 2018).

22. Lisa McTigue Pierce, "The Sustainability of Ecommerce Packaging Is in Question," *Packaging Digest,* February 21, 2017, http://www.packaging digest.com/sustainable-packaging/the-sustainability-of-ecommerce -packaging-is-in-question-2017-02-21 (accessed June 13, 2018).

23. Lisa McTigue Pierce, "Reusable…Refillable…Remarkable," *Packaging Digest,* January 31, 2013, http://www.packagingdigest.com

/smart-packaging/reusablerefillableremarkable (accessed June 13, 2018).

24. Replenish, http://myreplenish.com (accessed May 21, 2018).

25. Be Green Packaging, begreenpackaging.com (accessed May 21, 2018).

26. BioCycle, http://www.findacomposter.com (accessed May 10, 2018).

27. Lisa McTigue Pierce, "Walmart Unveils New Sustainable Packaging Priorities," October 26, 2016, *Packaging Digest,* http://www.packaging digest.com/sustainable-packaging/walmart-unveils-new-sustainable -packaging-priorities-2016-10-26 (accessed May 22, 2018).

28. "Sustainable Packaging," *Packaging Digest,* http://www.packaging digest.com/sustainable-packaging (accessed May 22, 2018).

29. "What's Your Score?" Wal-Mart's Sustainable Packaging Scorecard, 2008, http://www.sustainability-education.com (accessed May 22, 2018).

CHAPTER 10

The Forgotten Ones: Pre-consumer Waste

1. Unilever, "Unilever Commits to 100% Recyclable Plastic Packaging by 2025" (news release), January 14, 2017, https://www.unilever .com/news/press-releases/2017/Unilever-commits-to-100-percent -recyclable-plastic.html (accessed May 10, 2018).

2. The scope was set to cover nonhazardous waste only. This accounts for 99.9 percent of the total waste and ensured that Unilever met strict hazardous waste disposal requirements specific to the countries in which it was operating.

3. Gina-Marie Cheeseman, "The Journey to Corporate Zero-Waste-to-Landfill," Triple Pundit, April 3, 2017, https://www.triplepundit.com/ special/waste-management-covanta/journey-zero-waste-landfill (accessed May 22, 2018).

4. Andrew Burger, "Zero-Waste-to-Landfill Gets Certified," Triple Pundit, August 20, 2014, https://www.triplepundit.com/special /setting-the-standard/zero-waste-landfill-gets-certified (accessed May 22, 2018).

5. Panasonic, "Environment: Factory Waste Management—Zero Waste Emissions," https://www.panasonic.com/global/corporate /sustainability/eco/resource/zero.html (accessed May 22, 2018).

6. Panasonic, "Environment: Factory Waste Management."

7. Plastic Forests, http://plasticforests.com.au (accessed May 10, 2018).

8. Jessica Lyons Hardcastle, "Unilever Chief Supply Chain Officer on How to Achieve Zero Waste," Environmental Leader, February 15, 2016, https://www.environmentalleader.com/2016/02/unilever-chief -supply-chain-office-on-how-to-achieve-zero-waste (accessed May 10, 2018).

9. Sharon Guynup, "The Zero-Waste Factory," *Scientific American,* July 13, 2017, https://www.scientificamerican.com/custom-media /scjohnson-transparent-by-design/zerowastefactory (accessed May 10, 2018).

10. Ellen MacArthur Foundation, "Circular Economy Overview," https:// www.ellenmacarthurfoundation.org/circular-economy/overview /concept (accessed May 10, 2018).

11. Unilever, "Unilever: The Zero Waste Mindset," published July 2, 2015, https://www.youtube.com/watch?v=cEb-yZy3mZM (accessed May 10, 2018).

12. Unilever, "Zero Waste to Landfill across Unilever's Global Factory Network," published April 24, 2015, https://www.youtube.com /watch?v=W700bpAPdQw (accessed May 10, 2018).

13. Joann Muller, "How GM Makes $1 Billion a Year by Recycling Waste," *Forbes,* February 21, 2013, https://www.forbes.com/sites /joannmuller/2013/02/21/how-gm-makes-1-billion-a-year-by -recycling-waste/#355b04302309 (accessed May 10, 2018).

14. Muller, "How GM Makes $1 Billion a Year."

15. Muller, "How GM Makes $1 Billion a Year."

16. Veena Sahajwalla, "Micro-Factories Are Home-Grown Answer to Incredible Rubbish Recycling Problem," *Sydney Morning Herald,* February 28, 2018, https://www.smh.com.au/national/micro-factories -are-home-grown-answer-to-incredible-rubbish-recycling-problem -20180228-p4z25w.html (accessed May 10, 2018).

CHAPTER 11

Consumers Care

1. Dictionary.com, s.v., "Consumer," http://www.dictionary.com/browse /consumer (accessed May 21, 2018).

2. KoAnn Skrzyniarz, Carol Cone, Wendy Salomon, et al., "Enabling the Good Life: A Sustainable Brands' Research Study in Partnership with Harris Poll," 2017, https://insights.sustainablebrands.com/full-report (accessed May 22, 2018).

3. Skrzyniarz et al., "Enabling the Good Life."

4. Skrzyniarz et al., "Enabling the Good Life."

5. Lush USA, "Our Values: Naked," https://www.lushusa.com/Stories -Article?cid=article_our-values-naked (accessed May 21, 2018).

6. Lush USA, "Recycled Black Pots," https://www.lushusa.com/Stories -Article?cid=article_recycled-black-pot (accessed May 21, 2018).

7. Lush USA, "Our Values: Naked."

8. Lush USA, "Innovation vs Globalization," https://www.lushusa.com /Stories-Article?cid=article_innovation-vs-globalization (accessed May 22, 2018).

9. Lush USA, "Charitable Giving," https://www.lushusa.com/charitypot .html (accessed May 21, 2018).

10. Dr. Bronner's Magic Soaps, "150 Years and 5 Generations of Family Soapmaking," https://www.drbronner.com/about/ourselves/the-dr -bronners-story (accessed May 21, 2018).

11. Organic Consumers Association, "Dr. Bronner's Magic Soaps Files Lawsuit against Major 'Organic' Cheater Brands" (news release), April 28, 2007, https://www.organicconsumers.org/press/dr-bronners -magic-soaps-files-lawsuit-against-major-organic-cheater-brands (accessed May 21, 2018).

12. Lisa Bronner, "Dilutions Cheat Sheet for Dr. Bronner's Pure-Castile Soap," June 21, 2017, https://www.drbronner.com/all-one -blog/2017/06/dilutions-cheat-sheet-dr-bronners-pure-castile -soap (accessed May 21, 2018).

13. Dr. Bronner's Magic Soaps, "100% Post-Consumer Recycled Plastic Packaging—Bottles into Bottles, Not Landfill!" https://www.drbronner .com/about/our-earth/packaging-innovations (accessed May 21, 2018).

14. Dr. Bronner's, "100% Post-Consumer Recycled."

15. Dr. Bronner's, "100% Post-Consumer Recycled."

16. Jenna Igneri, "These Are the Sustainable Retailers You Should Be Buying From: Bringing Eco-Friendly to the Masses," Nylon, April 19,

2017, https://nylon.com/articles/sustainable-retailers-earth-day (accessed May 22, 2018).

17. Nielsen, "It Pays to Be Green: Corporate Social Responsibility Meets the Bottom Line" (news release), June 17, 2014, http://www.nielsen .com/us/en/insights/news/2014/it-pays-to-be-green-corporate -social-responsibility-meets-the-bottom-line.html?afflt=ntrt 15340001&afflt_uid=WM1VXo82qYs.ut0ZtHw37dTO4y_cyjJztf8aE aarw5i5&afflt_uid_2=AFFLT_ID_2 (accessed May 21, 2018).

18. Jenny Purt, "Consumer Behaviour and Sustainability—What You Need to Know," *The Guardian,* September 10, 2014, https://www .theguardian.com/sustainable-business/2014/sep/10/consumer -behaviour-sustainability-business (accessed May 21, 2018).

CHAPTER 12
Designing for the New Consumer: Abundance without Waste

1. "The Five Human Aspirations," BBMG, October 10, 2016, http:// bbmg.com/the-five-human-aspirations (accessed May 10, 2018).

2. Victor Reklaitis, "Why the 'Sharing Economy' Doesn't Work for International Package Delivery—Yet," MarketWatch, September 12, 2015, https://www.marketwatch.com/story/why-the-sharing-economy -doesnt-work-for-international-package-delivery-yet-2015-09-03 (accessed May 22, 2018).

3. Statista, "Number of Active Etsy Buyers from 2012 to 2017 (in 1,000)," 2018, https://www.statista.com/statistics/409375/etsy-active-buyers (accessed May 22, 2018).

4. Wikipedia, s.v. "Airbnb," last modified June 10, 2018, 20:32, https:// en.wikipedia.org/wiki/Airbnb.

5. JP Mangalindan, "Meet Airbnb's Hospitality Guru," *Fortune,* November 20, 2014, http://fortune.com/2014/11/20/meet-airbnb-hospitality -guru (accessed May 22, 2018).

6. Sarah Lacy, "How Daniel Became Goliath," Startups.co, March 12, 2017, https://www.startups.co/articles/how-daniel-became-goliath (accessed May 22, 2018).

7. Arthur Friedman, "These Companies Are Making Strides to Extend Clothing Life to Curb Textile Waste," *Sourcing Journal,* October 31, 2017, https://sourcingjournalonline.com/textile-recycling-evolving -repair-reuse-extend-clothing-life (accessed May 10, 2018).

8. Anya Khalamayzer, "Eileen Fisher Has Designs on Keeping Clothing out of Landfills," Green Biz, November 21, 2016, https://www.greenbiz .com/article/eileen-fisher-has-designs-keeping-clothing-out-landfills (accessed May 10, 2018).

CHAPTER **13**

Changing the Paradigm to Enable and Inspire Responsible Consumption

1. Virginie Helias, "Driving Positive Consumption in Pursuit of the Good Life," LinkedIn, June 6, 2017, https://www.linkedin.com/pulse /driving-positive-consumption-pursuit-good-life-virginie-helias ?articleId=6276102268441886720 (accessed May 10, 2018).

2. Virginie Helias, "Why the Future of Consumption Is Circular," World Economic Forum, January 15, 2018, https://www.weforum.org /agenda/2018/01/future-consumption-circular-economy-sustainable (accessed May 10, 2018).

3. "P&G and Gruppo Angelini's JV Fater Develops New Technology to Recycle 10,000 Tonnes/Yr of Used Diapers into New Products and Materials," RISI Technology Channels, October 31, 2017, https:// technology.risiinfo.com/tissue/europe/pg-and-gruppo-angelinis-jv -fater-develops-new-technology-recycle-10000-tonnesyr-used -diapers-new-products-and-materials (accessed May 10, 2018).

4. Procter & Gamble, "P&G's Head & Shoulders Creates World's First Recyclable Shampoo Bottle Made with Beach Plastic" (news release), January 19, 2017, http://news.pg.com/press-release/head-shoulders /pgs-head-shoulders-creates-worlds-first-recyclable-shampoo-bottle -made- (accessed June 13, 2018).

5. United Nations Climate Change, "Winners of 2017 UN Climate Solutions Award Announced" (news release), October 12, 2017, http:// newsroom.unfccc.int/climate-action/winners-of-2017-un-climate -solutions-awards-announced (accessed May 23, 2018).

6. "Is Laundry Only a Woman's Job? A Nielsen India Survey," Procter & Gamble, January 16, 2015, https://www.rewardme.in/home/home -upkeep/article/ariel-india-movement (accessed May 10, 2018).

7. M&M Global Staff, "How Ariel's 'Share the Load' Campaign Conquered International Media Awards," M&M Global, January 10, 2017, http://mandmglobal.com/how-ariels-share-the-load-campaign -conquered-international-media-awards (accessed May 10, 2018).

8. Sheryl Sandberg, "This is one of the most powerful videos I have ever seen—showing how stereotypes hurt all of us and are passed from generation to generation," Facebook video, February 24, 2016, https://www.facebook.com/sheryl/videos/vb.717545176/10156510941810177/?type=2&theater (accessed May 10, 2018).

CHAPTER 14
Value for Business in the Circular Economy

1. European Parliament, "Strategy on Plastics in the Circular Economy," http://www.europarl.europa.eu/legislative-train/theme-new-boost-for-jobs-growth-and-investment/file-strategy-on-plastics-in-the-circular-economy (accessed May 10, 2018).

2. Fern Shen, "Howard County to Expand Recycling to Include Plastics," October 25, 1989, https://www.washingtonpost.com/archive/local/1989/10/25/howard-county-to-expand-recycling-to-include-plastics/3feae094-ff9c-4b5c-89b7-b4a99e06d7b0/?utm_term=.0ec34f444bab (accessed May 10, 2018).

3. Nespresso, "The Positive Cup," https://www.nespresso.com/positive/us/en#!/sustainability (accessed May 10, 2018).

4. Keurig Recycling, www.keurigrecycling.com (accessed May 10, 2018).

Acknowledgments

THIS BOOK WOULD NOT HAVE BEEN POSSIBLE WITHOUT the hard work of many people. First, thanks to Veronica Rajadnya and Lauren Taylor, who spent countless hours in conversations and writing and refining content. Their effort is immeasurable.

Next, thanks to Laurent Gerbet, Megan Byers, Amanda Nicholson, Heather Grizzle, Sarah Schaefer, Matt Demorais, and Leslie Isaac, who provided support to make sure chapters were finished on time.

Of course, a big thank-you to Steve Piersanti at Berrett-Koehler for his guidance and support of this project from start to finish.

Even though they can't be named in person, we want to acknowledge all the people out there every day advocating for environmental change, whether in their communities, workplaces, government, or globally. And to the next generation of packaging designers and sustainability leaders: you are part of the solution.

Index

Page locators in *italics* represent figures and tables.

About the Author

TOM SZAKY IS FOUNDER AND CEO OF TerraCycle, a global leader in the collection and repurposing of hard-to-recycle waste. TerraCycle operates in 21 countries, working with some of the world's largest brands, retailers, cities, and manufacturers to establish national platforms to recycle products and packaging that currently go to landfill or incineration.

Through TerraCycle, Tom is pioneering new waste management processes involving manufacturers and consumers to devise circular solutions for hundreds of waste streams—including cigarette butts, laboratory waste, used coffee capsules, dirty diapers, used chewing gum, old plastic gloves, and even flexible food packaging—that otherwise have no path to be recycled. As one example, TerraCycle developed and operates the largest supply chain for ocean plastic in the world, partnering with such companies as Procter & Gamble to integrate this material into their product packages.

Tom and TerraCycle have received hundreds of social, environmental, and business awards, as well as recognition from a range of organizations, including the United Nations, the US Chamber of Commerce, *Fortune* magazine, the World Economic Forum, and the Schwab Foundation.

Tom is the author of three other books, *Revolution in a Bottle* (2009), *Outsmart Waste* (2014), and *Make Garbage Great* (2015). He created, produced, and starred in TerraCycle's reality show, *Human Resources*, which aired three 10-episode seasons on Pivot from 2014 to 2016 and is now syndicated in more than 20 foreign markets; it is also available to stream on Amazon and iTunes. Tom also writes for news outlets, from the *New York Times* and TreeHugger to HuffPost and *Entrepreneur* magazine.

Tom serves on the boards of directors of the Product Stewardship Institute, D'Addario Foundation, World Economic Forum (Future of Consumption board), and Ellen McArthur Foundation (CE100 board). He holds an international baccalaureate degree from Upper Canada College and attended Princeton University, where he developed the idea for TerraCycle. He is a frequent global lecturer, including annual appearances at Princeton University, the Wharton School of Business, Harvard Business School, and Yale University.

About TerraCycle

TERRACYCLE IS THE WORLD'S LEADER IN the collection and repurposing of complex waste streams, ranging from used cigarette butts and dirty diapers to coffee capsules and ocean plastic. It was incorporated in 2003 as the manufacturer of a simple organic fertilizer sold in used soda bottles collected from recycling bins, creating the world's first product made from *and* packaged entirely in waste.

When TerraCycle's founder and CEO, Tom Szaky, realized he could make a bigger impact in the world by using product and packaging waste, deemed to have no value, to create raw materials sold to manufacturers to produce new products, he changed the business model. TerraCycle no longer makes the fertilizer and is now an international leader in recycling the unrecyclable.

Across 21 countries, TerraCycle partners with major consumer goods manufacturers—such as Procter & Gamble, Colgate-Palmolive, L'Oréal, and many more—to run free collection programs that, in return for the waste, make a small donation to a school or charity. More than 4 billion pieces of pre- and post-consumer packaging have been collected and almost $22 million has been donated to schools and nonprofits.

TerraCycle launched the world's first recycling program for cigarette butts and recycles hundreds of millions of them around the world. Through its Zero Waste Boxes, TerraCycle enables individuals and groups to recycle hundreds of waste streams in their homes and businesses.

In 2017 TerraCycle, Procter & Gamble, and SUEZ made the world's first recyclable shampoo bottle made from plastic collected from beaches and waterways. TerraCycle has also acquired Air Cycle Corporation, giving TerraCycle its first foray into waste disposal mandated by federal regulations.

TerraCycle and Tom Szaky have received hundreds of social, environmental, and business awards, as well as recognition from a range of organizations.

Also by Tom Szaky

Outsmart Waste

The Modern Idea of Garbage and How to Think Our Way Out of It

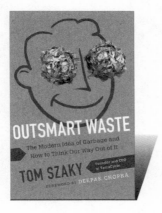

Ever-expanding landfills, ocean gyres filled with floating plastic mush, endangered wildlife. Our garbage has become a massive and exponentially growing problem in modern society. Eco-entrepreneur Tom Szaky explores why this crisis exists and explains how we can solve it by eliminating the very idea of garbage. To outsmart waste, he says, we first have to understand it, then change how we create it, and finally rethink what we do with it.

By mimicking nature and focusing on the value inherent in our by-products, we can transform the waste we can't avoid creating from useless trash to a useful resource. Szaky demonstrates that there is value in every kind of garbage, from used chewing gum to juice pouches to cigarette butts. After reading this mind-expanding book, you will never think about garbage the same way again.

Paperback, 168 pages, ISBN 978-1-62656-024-6
PDF ebook, ISBN 978-1-62656-025-3
ePub ebook ISBN 978-1-62656-026-0

BK Berrett–Koehler Publishers, Inc.
www.bkconnection.com   **800.929.2929**

Berrett–Koehler
BK Publishers

Berrett-Koehler is an independent publisher dedicated to an ambitious mission: *Connecting people and ideas to create a world that works for all.*

Our publications span many formats, including print, digital, audio, and video. We also offer online resources, training, and gatherings. And we will continue expanding our products and services to advance our mission.

We believe that the solutions to the world's problems will come from all of us, working at all levels: in our society, in our organizations, and in our own lives. Our publications and resources offer pathways to creating a more just, equitable, and sustainable society. They help people make their organizations more humane, democratic, diverse, and effective (and we don't think there's any contradiction there). And they guide people in creating positive change in their own lives and aligning their personal practices with their aspirations for a better world.

And we strive to practice what we preach through what we call "The BK Way." At the core of this approach is *stewardship,* a deep sense of responsibility to administer the company for the benefit of all of our stakeholder groups, including authors, customers, employees, investors, service providers, sales partners, and the communities and environment around us. Everything we do is built around stewardship and our other core values of *quality, partnership, inclusion,* and *sustainability.*

This is why Berrett-Koehler is the first book publishing company to be both a B Corporation (a rigorous certification) and a benefit corporation (a for-profit legal status), which together require us to adhere to the highest standards for corporate, social, and environmental performance. And it is why we have instituted many pioneering practices (which you can learn about at www.bkconnection.com), including the Berrett-Koehler Constitution, the Bill of Rights and Responsibilities for BK Authors, and our unique Author Days.

We are grateful to our readers, authors, and other friends who are supporting our mission. We ask you to share with us examples of how BK publications and resources are making a difference in your lives, organizations, and communities at www.bkconnection.com/impact.

Dear reader,

Thank you for picking up this book and welcome to the worldwide BK community! You're joining a special group of people who have come together to create positive change in their lives, organizations, and communities.

What's BK all about?

Our mission is to connect people and ideas to create a world that works for all.

Why? Our communities, organizations, and lives get bogged down by old paradigms of self-interest, exclusion, hierarchy, and privilege. But we believe that can change. That's why we seek the leading experts on these challenges—and share their actionable ideas with you.

A welcome gift

To help you get started, we'd like to offer you a **free copy** of one of our bestselling ebooks:

www.bkconnection.com/welcome

When you claim your **free ebook**, you'll also be subscribed to our blog.

Our freshest insights

Access the best new tools and ideas for leaders at all levels on our blog at ideas.bkconnection.com.

Sincerely,

Your friends at Berrett-Koehler